Stavros Kromidas

More Practical Problem Solving in HPLC

Further Titles of Interest

S. Kromidas

Practical Problem Solving in HPLC

2000, ISBN 3-527-29842-8

P.C. Sadek

The HPLC Solvent Guide

2002, ISBN 0-471-41138-8

P.C. Sadek

Troubleshooting HPLC Systems
A Bench Manual

1999, ISBN 0-471-17834-9

Stavros Kromidas

More Practical Problem Solving in HPLC

with Contributions by
Friedrich Mandel, Jürgen Maier-Rosenkranz and Hans-Joachim Kuss

Translated by
Renate FitzRoy

WILEY-VCH Verlag GmbH & Co. KGaA

Dr. Stavros Kromidas
Rosenstraße 16
66125 Saarbrücken
Germany
info@kromidas.de

Dr. Friedrich Mandel
Senior Applications Chemist
Mass Spectrometry
Agilent Technologies
Sales & Services GmbH & Co. KG
Hewlett-Packard-Str. 8
76337 Waldbronn
Germany
friedrich_mandel@agilent.com

Dr. Juergen Maier-Rosenkranz
GROM Chromatography GmbH
a GRACE Vydac Division
Etzwiesenstraße 37
72108 Rottenburg-Hailfingen
Germany
Juergen.Maier-Rosenkranz@grace.com

Dr. Hans-Joachim Kuss
Innenstadtklinikum der LMU
Nussbaumstr. 7
80338 München
Germany
Hans-Joachim.Kuss@med.uni-muenchen.de

This book was carefully produced. Nevertheless, authors and publisher do not warrant the information contained therein to be free of errors. Readers are advised to keep in mind that statements, data, illustrations, procedural details or other items may inadvertently be inaccurate.

Library of Congress Card No.: applied for

A catalogue record for this book is available from the British Library.

Bibliographic information published by Die Deutsche Bibliothek
Die Deutsche Bibliothek lists this publication in the Deutsche Nationalbibliografie; detailed bibliographic data is available in the Internet at http://dnb.ddb.de

© 2005 WILEY-VCH Verlag GmbH & Co. KGaA, Weinheim
All rights reserved (including those of translation in other languages). No part of this book may be reproduced in any form – by photoprinting, microfilm, or any other means – nor transmitted or translated into machine language without written permission from the publishers. Registered names, trademarks, etc. used in this book, even when not specifically marked as such, are not to be considered unprotected by law.

Printed in the Federal Republic of Germany.
Printed on acid-free paper.
Typesetting K+V Fotosatz, Beerfelden
Printing betz-druck gmbh, Darmstadt
Bookbinding J. Schäffer GmbH & Co. KG, Grünstadt

ISBN 3-527-31113-0

To my wife

Foreword

Over the last 35 years, HPLC has become the analytical separation method par excellence. HPLC instruments are standard equipment in analytical laboratories, in third place after scales and pH meters. Many introductions, compendia and textbooks have been written on the subject of HPLC that give more or less systematic description of the basic apparatus, various techniques and quantitative evaluation of chromatograms. All these books require systematic study – at least of some individual chapters.

This book, however, uses a different, sometimes quite idiosyncratic approach to HPLC. It provides practical support – answering questions of the "what do I do if..." variety. As even minute and often inadvertent changes in the HPLC system can cause heretofore-successful separations to go awry – e.g. a different supplier of solvents or chemicals, subtle changes (volumetric measurements at different temperatures) in the composition of eluents etc. – this book is an antidote to potential frustration. Over 90 tips deal with the choice of column, problems with buffers and eluent composition, troubleshooting etc. giving the individual users support in their daily routine. The author can build on his vast experience in HPLC.

I hope that his slightly unconventional description of HPLC technique will help many users to cope with their frustration with badly documented analytic systems. Perhaps, some of you may even feel inspired to document not only the process (drying at 40 °C), but also the performance (drying at 40 °C until the weight remains constant), and keep a record of chromatographic parameters for the most important analytes or those most difficult to separate.

June 2003   Prof. Dr. Dr. h.c. Heinz Engelhardt

Contents

Preface XV

The Structure of the Book 1

Part 1 (general section)   1
Part 2 (specific questions)   1

In Lieu of an Introduction 3

Chromatography Crossword   4
Across   4
Down   4
An HPLC-Quiz   6
An HPLC Tale   9
The Tale of Peaky and Chromy   9

1   HPLC Tips   11

1.1   Stationary Phases and Columns   11

Tip No.
01   "It improves with age" is a rule that applies to port and sometimes
          to red wine, but how about your C_{18} column?   11
02   Optimization via column parameters – what works best?   14
03   Can selectivity always be put down to chemical interactions
          with the stationary phase?   17
04   A matter of perspective ... Or: Selectivity and peak symmetry
          of basic compounds using reversed-phase packing materials   19
05   Separation of isomers   21
06   When should I use a "polar" C_{18} phase?   23
07   Are polar RP-C_{18} phases more suitable for the separation of polar analytes
          than non-polar phases?   24
08   What about non-endcapped phases – are they a thing of the past?   25
09   How can I separate acids using RP C_{18}?   27
10   The nitrile phase – some like it polar   29
11   The selectivity of RP columns   31

1.2   Buffers, pH Value   33

12   Does it always have to be potassium phosphate?   33

Tip No.	
13	UV cut-off of buffer solutions 34
14	Sources of errors when using buffers 35
15	The drawbacks of using buffers 37
16	Why is the pH value so important, and what does it do? 40
17	Why does the pH value shift even though I am using the correct buffer and the buffer capacity is sufficient? 42
18	Changes to the pH value in the eluent: the extent of the shift and the reasons behind it 43
19	An unintentional pH shift and its consequences 46
20	RP separations in the alkaline medium 49
21	Separation of basic and acidic compounds contained in the same sample 51

1.3 Optimization, Peak Homogeneity   53

22	The peaks appear too soon – what can be done? 53
23	What can I do if the peaks elute late? 55
24	Quick optimization of an existing gradient method 60
25	Increasing efficiency – often the fast track to success 63
26	Additives to the eluent 66
27	Separating the unknown – where shall I begin? 69
28	Separation of an unknown sample using a reversed-phase C_{18} column – how do I go about it? 72
29	Developing an RP separation – the two-day-method Part 1: Choice of column and eluent 74
30	Developing an RP separation – the two-day method Part 2: Fine-tuning of the separation 78
31	Quick check on peak homogeneity – Part 1 80
32	Quick check on peak homogeneity – Part 2 82
33	Tied to a standard operating procedure – how can a bad separation be improved further? 84
34	More elaborate measures to check peak homogeneity 86
35	First easily digestible tip 91
36	Second easily digestible tip 94
37	Third easily digestible tip 96

1.4 Troubleshooting   99

38	How to approach problems in a systematic manner 99
39	Spikes in the chromatogram 101
40	Additional peaks in trace analysis separations 103
41	What causes a ghost peak? 105
42	Ghost peaks in a blank gradient 107
43	Strange behaviour of a peak. What could be the cause? 108
44	When could one expect a change in the elution order of the peaks? 110
45	Tailing in RP HPLC – Part 1: Fast troubleshooting 114
46	Tailing in RP HPLC – Part 2: Further causes and time-served cures 116

Tip No.

47	Peak deformation and a shift in retention time due to an unsuitable sample solvent	119
48	Is flushing with water or acetonitrile sufficient?	123
49	Flushing and washing fluids for HPLC apparatus	125
50	When does the peak area change?	127
51	Reasons for a change in either peak height or peak area, but not in both	129
52	Excesses and their pitfalls	131
53	Algae, fungi and bacteria in HPLC	132
54	Does 40 °C always mean 40 °C?	134
55	The most common reason for a lack of reproducibility is a lack of methodological robustness	135

Have a break ... 138

Dear Reader 138

Complete the sentences 139

"Matching pairs" 140

Has Peaky remembered his lessons correctly? 141

1.5 General HPLC Tips 142

56	What changes can you expect when switching from one HPLC system to another?	142
57	What changes can be expected in a chromatogram if the dead volume is larger in one isocratic system than in another?	144
58	Contribution of the individual modules of the system to band broadening	146
59	How to keep retention times constant while reducing the diameter of the column	148
60	Has 3 µm material been developed sufficiently to be used in routine separations?	150
61	Miniaturization may be all well and good – but when does it really work and does it make sense in routine separations?	152
62	Why is it that peaks appear later with a new column?	154
63	Column length, flow and retention times in gradient separations	155
64	Column dimensions and gradient separations	159
65	What is the difference between dead time and dead volume on the one hand and selectivity and resolution on the other?	161
66	Troublesome small peaks	163
67	Lowering the detection limit by optimizing the injection	164
68	Setting the parameters of an HPLC instrument	167
69	The right wavelength – old hat to some, a revelation to others	171
70	Characteristics of refraction, fluorescence and conductivity detectors	175

Tip No.
71	Does it always have to be HPLC?	177
72	Methanol versus acetonitrile	180
73	Tale of a foursome pub-crawl – can peaks elute before the front?	183

References to HPLC-Tips   185

2   LC-MS Coupling, Micro- and Nano-LC, Quantification   187

2.1 by Friedrich Mandel, 2.2 by Jürgen Maier-Rosenkranz,
2.3.2–2.3.4 by Hans-Joachim Kuss

2.1   LC-MS Coupling   187

LC-MS – The one and only universal tool?   187

Tip No.
74	Choosing the right LC-MS interface	189
75	Which mobile phases are compatible with LC-MS?	195
76	Phosphate buffers – the exception	197
77	Paired ions	198
78	Using additives to enhance API-electrospray ionization	200
79	How can I enhance sensitivity of detection?	203
80	No linear response and poor dynamic range?	205
81	How much MS^n do I need?	208

Need more help?   210

References to LC-MS   211

Internet addresses of interest for LC-MS coupling   212

2.2   Micro- and Nano-LC   213

A short introduction   213

Tip No.
82	Lower efficiency – plate number too low Part 1: Effects of dead volume in the connecting parts. Which column diameter should be used with which capillary diameter?	214
83	Lower efficiency – plate number too low Part 2: Effects of injection amount and injection volume	215
84	Lower efficiency – plate number too low Part 3: Impact of flow cell (UV, fluorescence, radio detection)	216
85	No gain in sensitivity: flow cell – path length – S/N	218
86	Fused silica and PEEK capillary connections	219
87	Fast sample loading due to column switching	220

Tip No.		
88	Injection system: full loop injection, partial loop fill injection, timed programmed injection, direct injection	222
89	Protecting the system: cleaning-up guard column, saturation column	223
90	Retention time shift: gradient delay volume, mixing chamber volume, gradient accuracy	224
91	Transferability – downscaling: correct gradients	226

References to Micro- and Nano-LC   228

2.3 Quantification   229

2.3.1	Practical aspects of quantification in HPLC 229
2.3.1.1	Peak area or peak height? 229
2.3.1.2	What factors have an impact on the peak area? 230
2.3.1.3	Formulae and short statements or comments with respect to the quantification methods 231
2.3.1.4	Examples with actual figures 233
2.3.2	Quantification in Chromatography 236
2.3.2.1	Optimum separation – correct peak acquisition 236
2.3.2.2	Understanding the "mind" of the integration system 240
2.3.2.3	Setting parameters and their effect on peak area and peak height 241
2.3.2.4	Where can mistakes be made? 243
2.3.3	Methods of Quantification 244
2.3.3.1	What is the 100% method? 244
2.3.3.2	What is the external standard method? 244
2.3.3.3	Why use an internal standard? 245
2.3.3.4	In which cases should the additions method be used? 247
2.3.4	Weighted Regression 249
2.3.4.1	What about the F-test? What are the other possibilities? 249
2.3.4.2	How do we weight the individual values? 250
2.3.4.3	How to use Excel for weighted regression 250
2.3.5	Solutions to quantification problems 253

References   253

3 Appendix   261

3.1 Solutions to the Problems   261

3.1.1	Crossword – the solution 261
3.1.2	An HPLC quiz – the solution 261
3.1.3	An HPLC tale with Peaky and Chromy – the solution 264
3.1.4	Complete the sentences 265
3.1.5	"Matching pairs" 266
3.1.6	Did Peaky remember his lessons correctly? 267

3.2	**From Theory to Practice – Empirical Formulae, Rules of Thumb and Simple Correlations in Everyday HPLC** 269
3.3	**Information Resources for Analysis/HPLC** 277
3.4	**Analytical Chemistry Today** 281
3.5	**Trends in HPLC** 285
3.6	**Thoughts About a Dead Horse** 290

Index   291

Preface

The HPLC community gave "Practical Problem Solving in HPLC" a warm welcome.

Alongside joy, I also felt a kind of urge to "keep going". The logical result of this is "More Practical Problem Solving in HPLC". The intention, language and style have remained the same, serving one aim: The book is meant to be an easy-to-read companion for HPLC users, providing tips and suggestions in a compact form.

Alongside general tips we have also included three "Special Areas" in this volume. These are two techniques that are already important and will become increasingly so in future – LC-MS-coupling and micro-/nano-LC – as well as a look at quantitative evaluation. Even if today's computers do nearly all the work for us, the background could prove interesting for some readers, such as how settings influence the peak shape, area and height, or why the calculated content is dependent on the evaluation method used.

I would like to emphasize that the "Practical Problem Solving" series is not intended as a course book. Rather, it is a concise representation of the relations and explanations from a practical viewpoint. For the theoretical background I would point the reader towards the appropriate works.

I wish to extend my gratitude to my colleagues Friedrich Mandel, Joachim Maier-Rosenkranz and Hans-Joachim Kuss, who provided their expert knowledge in their specialized area.

The cooperation with Steffen Pauly at Wiley-VCH proved to be most pleasant. I also thank Renate FitzRoy for expertly translating the often not-trivial passages of the original manuscript into English, and Uwe Neue for his scientific discussions and critical reading of the text.

Finally, I hope you have fun while reading this book and that you find here ideas and help for your daily work with HPLC.

Saarbrücken, September 2004   Stavros Kromidas

The Structure of the Book

Part 1 (general section)

In the first part, I am trying to break the reader in gently before proceeding to the 73 tips in which various aspects of HPLC are discussed. Although it is not always possible to link everything to an overriding theme, I have tried to introduce the following subject categories:

- Stationary phases, columns (Tips Nos. 01–11)
- Buffers, pH value (Tips Nos. 12–22)
- Optimization, checking peak homogeneity (Tips Nos. 23–34)
- Troubleshooting (Tips Nos. 35–54)
- Miscellaneous tips (Tips Nos. 55–73)

In general, every tip is a self-contained unit discussing a specific problem, which means that the book does not have to be read from cover to cover. The reader can jump back and forth at leisure. However, a very important and complex subject may be spread over several tips, e.g., "Tailing in HPLC" is discussed in Tip Nos. 45 and 46.

Or the same problem may be discussed from different angles and crop up in two or three different tips, e.g., "sources of errors when using buffers" in Tip No. 14, and "Shift of pH value in the eluent" in Tip No. 18. What I am trying to achieve is to open up a variety of routes to the reader to make the most of these tips.

Where appropriate, references are given regarding tips that are related to the topic or provide additional information. For easier reference, the tips have been numbered. As some of you may already possess Volume 1 of the series "Practical Problem Solving in HPLC", I have also included it in my references. Whenever I refer to it, the figure 1 will appear behind a forward slash, e.g., Tip No. 34/1. If not stated otherwise, the chromatograms are results of my own measurements or they are examples from practical separation classes held at NOVIA GmbH, Frankfurt/Main to whom I would like to express my thanks.

Part 2 (specific questions)

Over recent years, many variants of classical HPLC as well as related separation techniques have been developed. The most important of these are in my opinion LC-MS coupling and micro- or nano-LC. Both have an important role to play in the future, which is why you will find tips referring to them in Part 2.

Finally, a word about quantification.

With the software programs that are now available, quantitative evaluation of chromatograms has become child's play. However, I thought it would perhaps be a good idea to give a brief overview of the integration and data handling methods, and the reader could draw some educational benefit from hands-on quantification using a range of methods. Both the pocket calculator and the personal computer approach are offered; the latter using MS Exal. This might even help to memorize and internalize these various methods. What we also wanted to achieve was to give some background to the integration process and demonstrate the impact of individual parameters on peak area and height to round off the discussion in Part 2.

The Appendix contains a bibliography, an index and further information on HPLC.

In Lieu of an Introduction

Dear Reader,

Now you hold this book in your hands and you may feel a little reluctant to jump in at the deep end and go straight to the serious subjects. If that's the case, take it easy and go through the fun pages first before you start on any earnest work. There is something for every taste.

1. Do you like a challenge? Have a go at the crossword on page 4.
2. Do you like solving riddles? There is a quiz waiting for you on page 7.
3. Are you a child at heart? Do you still enjoy being told stories? Then read the chromatographic tale of Peaky and Chromy on page 9.

You will find the answers from page 261 on.

Are you far too grown-up and serious to waste your time with childish games? All right, then go ahead and dive into the fountain of wisdom on page 11.

Chromatography – and more – Crossword

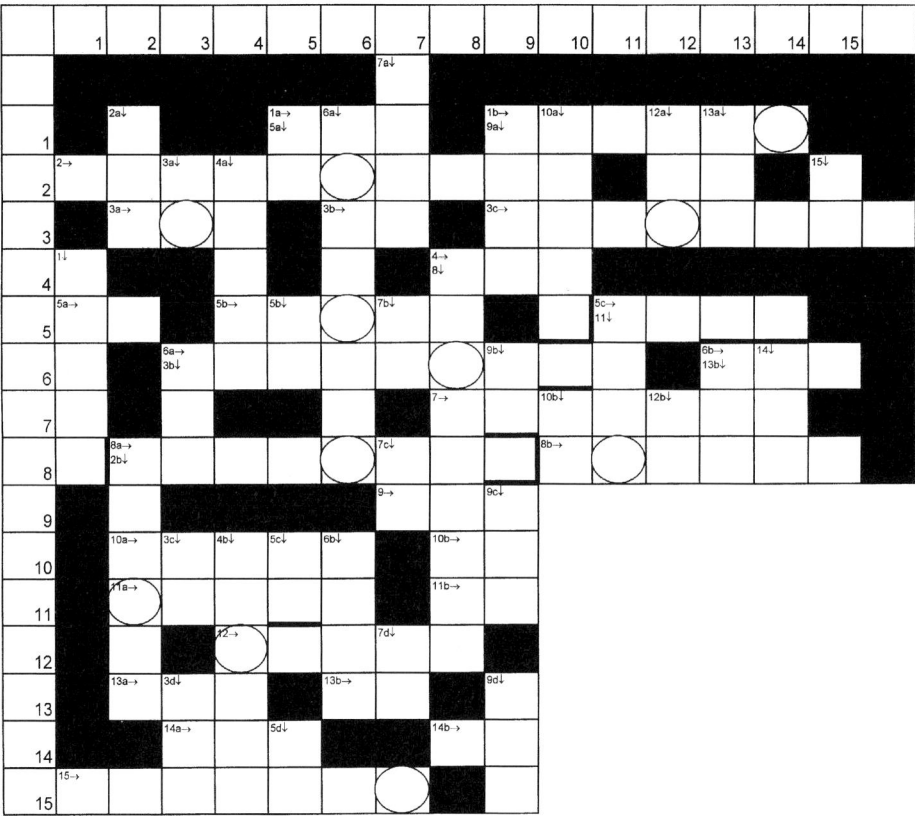

Across

1a	Whisky and port, but not necessarily columns improve with …
1b	Substance in the analyte of Tip No. 1
2	In normal life, it is measured on two different scales in the USA and Europe. In chromatography it can be crucial to the reproducibility of your results
3a	World Health Organization
3b	Acronym for Alaska
3c	Additive to an eluent
3d	United Nations
4	Repair
5a	Short for a Californian city
5b	What do you do with your pump in order to get rid of excess air?
5c	Used for hearing
5d	Summer time or Saint
6a	Depends on the interaction between sample and stationary phase
6b	Inevitable part of British school uniform
7a	Interrogative pronoun
7b	… of macromolecules from a C_{18} phase takes ages
8a	Phenomenon that occurs if a sample is not properly dissolved in the eluent
8b	Heading
9a	Non-interactive type of fitting, tubing and accessories used in HPLC
9b	Expressing your wish or opinion in an authorized formal way
10a	Mapping technique used in genetics
10b	Substance at one end of the pH spectrum
10c	Nothing, zero
11a	Help, support
11b	Cowboy competition
12a	French for 'wrong', harm
12b	Predator
13a	What runs through a column (plural)
14a	Turkish currency
14b	American news agency
15a	Woman with refined manners
15b	The method discussed in this book

Down

1a	Is usually kept constant throughout a separation
1b	In chromatography, it always is theoretical
2a	Not old!
2b	Chemical sign for iron
2c	Highly polar phase
3a	Initials of a Dutch housewife who became famous as a spy
3b	Noble gases are also called …
3c	Pagan
4a	Another name for hashish
4b	Animal you keep at home

4 c	All right
4 d	Blood vessel leading away from the heart
5 a	The first two of the five basic vowels
5 b	Abbreviation for retention time
5 c	Colloidal solution or Latin for sun
5 d	Chemical sign for nitrogen
5 e	Chemical sign for sodium
6 a	With this separation mode you can nearly always save time and always lower the limit of detection, but you can hardly ever improve selectivity
6 b	Make changes in a text, film or recorded piece of music
7 a	We like it narrow!
7 b	German column manufacturer with a US subsidiary in Easton PA
7 c	Electrically charged particle
7 d	Goddess of Dawn in Greek mythology
8 a	Just to underline its significance, here again is 6 a
8 b	Chemical symbol for aluminium
9 a	Essential part of lab equipment
9 b	Abbreviation for Illinois
9 c	Flexible polymer
9 d	Abbreviation for Reversed Phase Chromatography
10 a	Either ...
10 b	To put to some purpose
10 c	In and ...
11 a	Opposite of right
11 b	Greek for against
12 a	Preposition
12 b	Abbreviation for Information Technology
13 a	Chemical symbol for erbium
13 b	Preposition indicating a direction
14	Solid polymeric packing used in ion-exchange separations
15	What comes out of a column

When you put the letters with circles around them in the right order you will get something you want to achieve in HPLC.

Good luck!

An HPLC-Quiz

On the left, you will find the description of a situation. On the right, there is a list of possible answers or consequences. How many of these possible answers are correct? All, some, one or none?

The packing has deteriorated	K	The peaks appear later
	L	The peaks become broader
	E	Resolution declines
	M	Selectivity declines
	I	Tailing appears
The proportion of acetonitrile in the eluent is increased	S	The peak area changes
	T	The retention time decreases
	C	The peak height changes
	E	The plate number changes
	Y	The lifetime of the column increases
The temperature is increased (ordinary RP-system)	B	Selectivity improves
	Z	Resolution improves
	W	The plate number increases
	T	Efficiency improves
	P	The retention factor is increased
The flow rate is increased	C	The peak are increases
	F	Resolution improves
	H	The plate number increases
	J	Efficiency improves
	O	Selectivity decreases
Endcapped C_{18} phases	S	... provide better peak symmetry for bases
	Q	... achieve a better separation of strong acids
	T	... achieve a better separation of bases, but they are unsuitable for non-polar substances
	X	... are more stable in an acidic eluent than non-endcapped C_{18} phases
	F	... mean that the surface is absolutely non-polar
A conditioning or saturation column (column between pump and injector)	V	... saturates the eluent with silica gel and protects the analytical column
	A	... must contain material with the same particle size as the separation column

I	... raises the pressure
K	... must be filled with the same stationary phase as the separation column
C	... must also be thermostatically controlled in order to ensure the constant viscosity

The letters in front of the correct answers, put in the right order, will give you a thermodynamic factor that is a measure of the relative retention of two compounds. Its value is determined by the choice of stationary phase, the mobile phase composition and the temperature. Find the solution!

Happy puzzle-solving!

An HPLC Tale

The Tale of Peaky and Chromy

Once upon a time there were two peaks who were very good friends – little Peaky Acid and big Chromy Silicasky. Whenever they met up in the "The Last Drop" tavern after a long retention time, they usually had enough time to tell each other their latest adventures. Today it was the turn of the lively little chap Peaky:

- You know, we had really great fun when our friend Nicolas W. Pump – remember Pumpous Nick – wanted to separate me and the other strong boys of the Acid Gang. His boss, Mr. Chromadis, wanted to have us all quantified. Well, Nick took a 125 mm×4 mm endcapped Super-X-fantastic-pura pura C_{18} column and used an 85/15 (w/w) mixture of ACN/100 mmol phosphate buffer, pH=5 at a flow rate of 1 mL min^{-1}.

- And?

- Well, some of the others made their appearance after 1–2 minutes, while others took 4–5 minutes. He seemed to be quite happy. Using his software, he already had us measured.

- How is that?

- Just the usual things: height, area, asymmetry factor and theoretical plate number.

- And were you all tall and slender?

- No, two or three of us were on the small side, and they were carrying something that looked like a tail ...

- So there was some tailing.

- Yes, and because they were so small he couldn't really measure them, but that didn't seem to bother him.

- So was everything O.K.?

- No, he suddenly wanted us all to move towards the back. So he took a little more water, and we all came a little later. Our height and area changed until suddenly ...

- What happened?

- One of us appeared as a double peak. You know it was a very old column and the packing was past its best. But fortunately, Nick not only have his wits about him, he even had a second column in his cupboard!

- Did it at least work then?

- No, I don't think so. Anyhow, he started cursing and then soon went home. The next day ...

Peaky had no time to finish his sentence, as the two friends had to leave their cozy place and move on to the large cafeteria "The Dregs".

Is there anything you don't like about this story or is there something not quite logical about it? Perhaps good old Mr. Pump did not take the best decisions or could Peaky be wrong in places?

1 HPLC Tips

1.1 Stationary Phases and Columns

Tip No. 01 "It improves with age" is a rule that applies to port and sometimes to red wine, but how about your C_{18} column?

Problem/Question

Experience shows that in an HPLC column, quality declines over time and peaks tend to broaden. Has the opposite ever been observed?

Solution/Answer

Yes! Let us be clear that "deterioration of the C_{18}-column" can mean two things! Firstly, there is the mechanical wear and tear on the packing material, the extent of which depends on flow, temperature, number of injections and operating mode (isocratic or gradient). The decline in quality of the packing material manifests itself in broadened peaks and/or tailing, sometimes even in double peaks, while the retention time remains constant. Secondly, the stationary phase can undergo a qualitative change, e.g., if sample components are irreversibly adsorbed onto the surface of the stationary phase. This causes a shift in retention time as well as a change in selectivity. This second type of deterioration could also be a positive change.

We know that if non-endcapped or poorly endcapped phases with a large number of free, active silanol groups at the surface are used with basic compounds, they produce tailing peaks. This is not a pretty sight, and if more basic substances are injected over time, they may get stuck to the interfering silanol groups, blocking their activity. As a consequence, basic compounds in the current sample do not find free silanol groups to flirt with and are thus eluted earlier, producing neat symmetrical peaks. Figure 1-1 shows just one of many typical examples, the separation of phthalic acid, aniline and acetophenone using 70/30 (w/w) MeOH/H_2O eluent with a non-endcapped Resolve C_{18}-column.

As would be expected, on a new column, aniline (the last peak) produces considerable tailing. Some time ago, during an HPLC course, the same mixture was injected into a vintage 1984 Resolve column (see Figure 1-2).

During its lifetime, this column has probably seen so many basic substances that none of the silanol groups have survived. As a result, aniline finds nowhere to bind to and is eluted earlier, producing a symmetrical peak. Just recently, the same column has been put to the test again (see Figure 1-3).

The chromatogram of a mixture of phthalic acid, aniline, toluene and ethylbenzene looks very neat. On this ancient column, aniline (2nd peak) is eluted almost as symmetrically as on a modern base-deactivated column. Incidentally – just to make a practical point, this column has been dropped several hundred times on purpose. The

More Practical Problem Solving in HPLC. S. Kromidas
Copyright © 2005 WILEY-VCH Verlag GmbH & Co. KGaA, Weinheim
ISBN: 3-527-31113-0

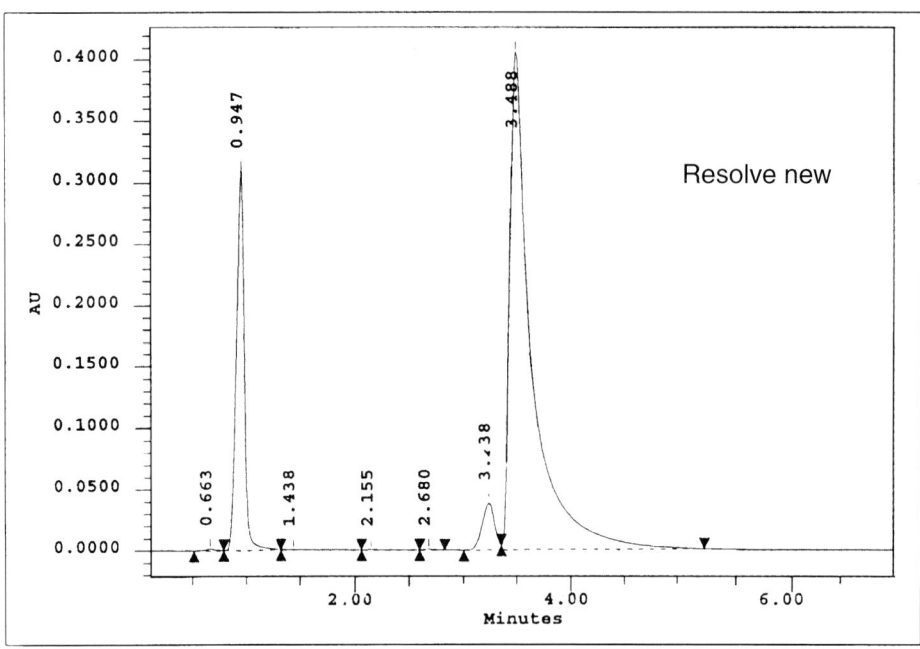

Figure 1-1. The separation of phthalic acid, aniline and acetophenone using a 70/30 (w/w) MeOH/H_2O eluent with a non-endcapped Resolve C_{18} column.

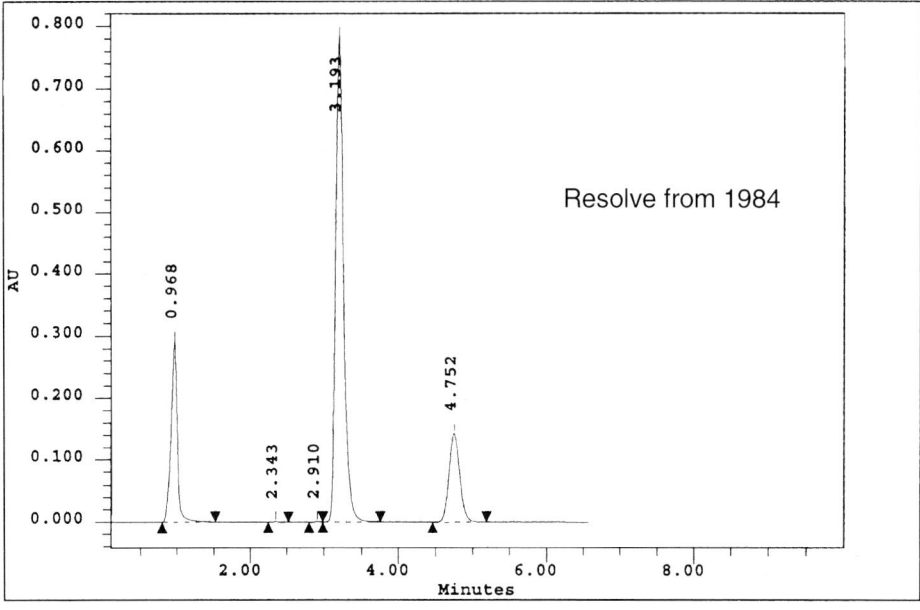

Figure 1-2. Separation of phthalic acid, aniline and acetophenone with a 70/30 (w/w) MeOH/H_2O eluent on a very old Resolve C_{18} column.

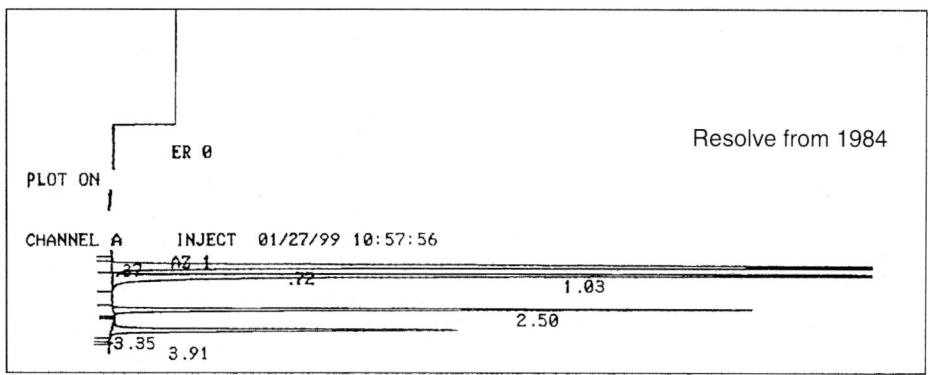

Figure 1-3. Separation of phthalic acid, aniline, toluene and ethylbenzene on the old (see Figure 1-2) Resolve column.

clear peaks of the other three components show that if a column, e.g., Resolve, is well packed, the packing material can easily survive such shock treatment.

Conclusion

If an HPLC phase irreversibly adsorbs problematic components, it may affect its properties, mainly in a negative way, but occasionally it may even turn out to be an improvement.

Tip No. 02 Optimization via column parameters – what works best?

Problem/Question

Suppose you want to use a certain stationary phase to perform an isocratic separation. It could be that your raw material supplier has used this material to validate the method, so you are stuck with it. Unfortunately, the first injection only produces a fairly lousy separation, and so does the second one. The equipment and everything else seem to be all right. Stationary and mobile phases being off limits, your boss gives you some leeway to experiment with the physical column characteristics and the flow. Thus, some of the parameters could be modified. The column can be lengthened, while the flow, particle size and inner diameter can be reduced. Whereas the first three measures will raise the number of theoretical plates, the last will reduce variance in the radial diffusion. Which of these options is the most effective?

Solution/Answer

Table 2-1 gives the resulting data (resolution, retention time, pressure) in relation to their physical parameters. The first row contains data from the first separation that was deemed insufficient (resolution $R_s = 1.1$) and marks our starting point.

First attempt

Reducing the flow to 0.5 mL min^{-1} reduces the pressure by a factor of two (from 45 bar to 22.5 bar) and increases the retention time by a factor of two (from 11 min to 22 min), but you will achieve a slight improvement in the resolution (1.3).

Second attempt

Lengthening the column by 1/3 (150 mm) slightly increases the retention time and the pressure while improving the resolution to 1.4.

Table 2-1. Physical properties of a column and the resulting chromatographic data (see text for further explanations).

Inner diameter (mm)	Particle size (µm)	Flow (mL min^{-1})	Length (mm)	Resolution	Retention time (min)	Pressure (bar)
4	5	1.0	100	1.1	11	45
4	5	0.5	100	1.3	22	22.5
4	5	1.0	150	1.4	16	68
4	5	1.0	250	1.9	27	113
2	5	0.28	100	1.1	11	ca. 45
4	3	1.0	100	1.6	11	126

Third attempt

The result of the second attempt may indicate that it would be worth trying out an even longer column. The best resolution could be achieved using a 250 mm column – but at a price! The pressure increased to 113 bar and the analysis time rose to 27 min.

Fourth attempt

Imagine you are reducing the inner diameter of the column from 4 mm to 2 mm, adjusting the flow to keep the linear velocity and thus the retention times constant. The increase in pressure (through reducing the inner diameter) and its reduction (through decreasing the flow) by a factor of 4 cancel each other out. Not only the pressure, but also the resolution will remain constant. What will change, however, is the peak height, which will increase if the same amount of sample is injected while the band-spreading in the column will be reduced.

Fifth attempt

Stick to the original column dimensions and the original flow but use 3 µm particles. While the analysis time remains constant at 11 min (owing to the column length and flow remaining constant) adequate resolution of 1.6 is obtained at a pressure of 126 bar.

Conclusion

1. Reducing the flow is easy to do. Unless the flow is reduced drastically, e.g., to 0.2 or 0.3 mL min^{-1}, this does not achieve very much (if you use small particles), and the drawback is an extremely long analysis time.
2. In isocratic separations where many peaks need to be separated and/or you are dealing with a complex matrix, the classical approach using a long column is still the best. Higher pressure and long analysis times are the downsides one has to put up with.
3. Reducing the inner diameter may not improve resolution, but it is a way of cutting down on eluent use and of lowering the detection limit (higher peaks!), which can be of some advantage when it comes to trace analysis and small samples.
4. If demands on peak capacity are not too extravagant and the samples are reasonably "clean", using small particles is often a sensible compromise – as long as the pressure remains acceptable.

Let us summarize

For isocratic separations:

- Matrix-loaded sample, high demands on selectivity? → Long column.
- Relatively "clean" sample? → 3 µm particles and a column length of 100 mm are adequate in most cases.

- Do you care about the environment? → Replacing your 4 mm columns with 3 mm ones reduces your solvent consumption by about 50% (!). In this day and age, you are unlikely to find an instrument that will not work with a 3 mm column anyway.

For gradient separations:

Column volume is not that critical (see Tips Nos. 63 and 64), so keep it small! For example, you can easily use a 75 mm or even shorter column.

Tip No. 03 — Can selectivity always be put down to chemical interactions with the stationary phase?

Problem/Question

We have all been taught that chromatographic separation results from interactions between the analytes and the stationary phase – with the exception perhaps of size exclusion chromatography. Depending on the mechanism, we presume that ionic or hydrophobic, or some other interactions, take place. Discussion of partitioning mechanisms has completely gone out of fashion. Are these interaction mechanisms the only things that are happening or is there anything else that has an impact on the selectivity of the chromatographic separation?

Solution/Answer

These interactions are not the only factors. Even for small molecules steric aspects can be important, see Figure 3-1. A separation of a mix of metabolites of tricyclic antidepressants using two "polar" RP-phases yields five peaks (centre, left). When using a material with a pore diameter of 300 Å six peaks appear. The desired selectivity is only achieved via an additional steric aspect introduced by the use of a packing material with a larger pore size.

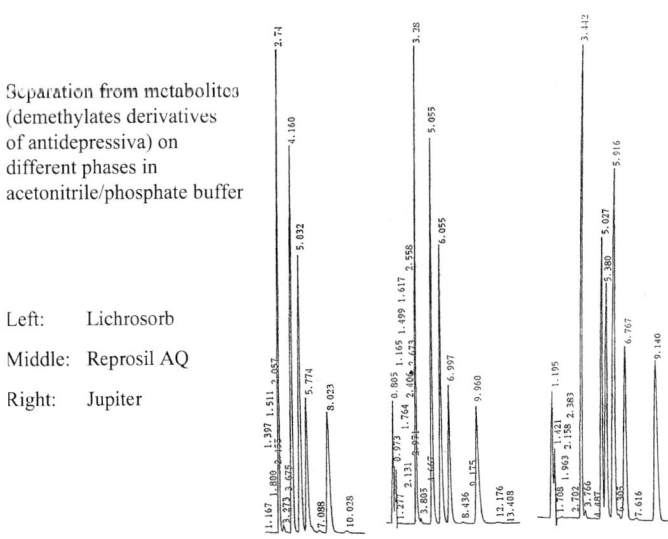

Separation from metabolites (demethylates derivatives of antidepressiva) on different phases in acetonitrile/phosphate buffer

Left: Lichrosorb
Middle: Reprosil AQ
Right: Jupiter

Figure 3-1.

Conclusion

Steric aspects cannot be excluded out of hand when separating an unknown sample. When developing a method and choosing a column, I therefore recommend considering phases with large and small pore diameters as well.

Examples of phases with a small pore diameter (60–80 Å):
NovaPak, Nucleosil 50, Spherisorb ODS 1/2 Superspher Select B.

Examples of phases with a large pore diameter (180–300 Å):
Symmetry 300, Jupiter, ProntoSIL 300, Zorbax SB 300, Discovery C_{18}.

Today it is popular to talk far too seriously about everyday matters.
It is about time somebody used everyday language for serious matters!

Tip No. 04

A matter of perspective...
Or: Selectivity and peak symmetry of basic compounds using reversed-phase packing materials

Problem/Question

Are you planning to separate strongly basic compounds? Well, well, well – be careful! Take, if you will, your favourite clean, expensive, state-of-the-art, metal-ion-free, super-endcapped phases with maximum surface coverage. Your peaks will be immaculate, sharp and symmetrical – the separation looks very neat, I will grant you that (see Figure 4-1).

But – are you sure you are getting the full picture? It pays to be circumspect when using polar/ionic components and hydrophobic stationary phases.

Solution/Answer

Just to be on the safe side, double-check your results – at least for the most important samples – using a polar RP phase and an unbuffered (!) methanol/water eluent. If you get the same number of naturally tailing peaks as you had using the hydrophobic phase, all is well. Don't be fooled by the sharpness of those peaks in Figure 4-1! It could be that this only gives the illusion of high selectivity. As Figure 4-2 shows, there should be four not just three peaks.

A phase with good surface coverage cannot produce polar/ionic interactions, so don't expect good polar/ionic selectivity! Be on your guard!

Your best bet is of course to find a polar phase that can produce polar interactions but that only has a small number of interfering silanol groups, such as SynergiPOLAR RP, Fluofix IEW or Zorbax SB C_8. These materials yield good selectivity, due to their polar character, as well as good peak symmetry, due to good coverage of the surface (see Figure 4-3).

Conclusion

If you are now convinced that you have to be vigilant when it comes to the separation of hydrophobic bases that's half the battle. Don't trust those sharp peaks – they may hide some crucial unresolved peak lurking in the background. However strict your supervisor, if you have actually checked peak homogeneity by using polar phases you will be able to prove that you have been taking extra care. This will earn you some brownie points in your laboratory. After all, your boss can't see everything, and as you don't (want to) know how he reacts if you overlook something, this might be a precaution worth taking.

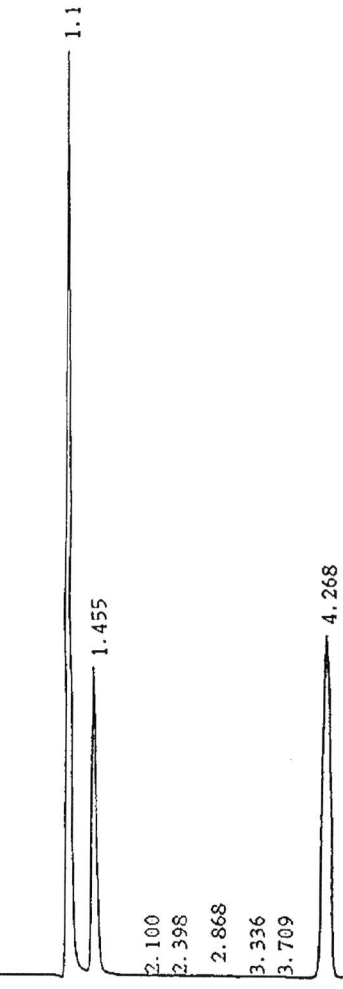

Figure 4-1. "Separation" of pyridine/benzylamine, uracil (inert!), phenol in an acidic phosphate buffer with a hydrophobic RP phase.

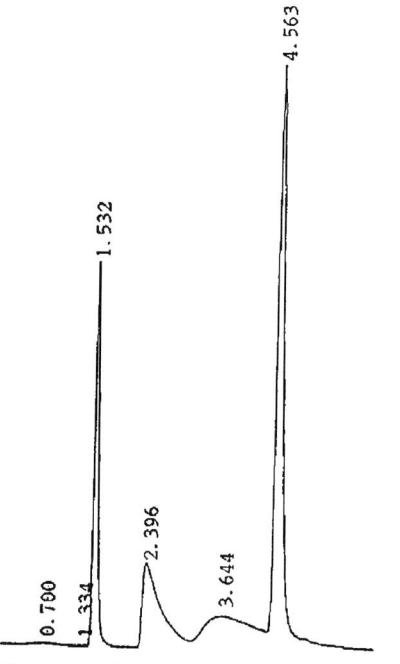

Figure 4-2. Separation of uracil (inert), pyridine, benzylamine, phenol in an acidic phosphate buffer with a polar RP phase.

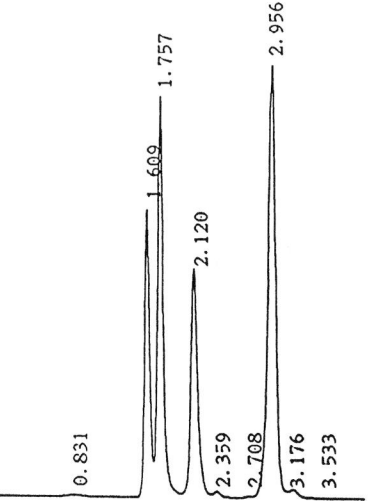

Figure 4-3. Separation of uracil (inert), pyridine, benzylamine, phenol in an acidic phosphate buffer with a Fluofix RP phase

Tip No. 05 — **Separation of isomers**

Problem/Question

The separation of isomers can prove to be quite difficult, as such compounds often have very similar properties. Which columns and eluents should be used in such cases?

Solution/Answer

Preliminary remark:
Difficult cases often call for unorthodox solutions. This general principle applies particularly when specific important problems such as customer complaints or acute toxicity issues need to be dealt with. We are not talking about developing a routine method where robustness and reproducibility of results are a major consideration. Keep this in mind when looking at the following suggestions!

Choice of column

Modern hydrophobic phases are hardly suitable for this problem. I recommend the following alternatives:

- **"Old classics" with polar functionality:**
  e.g., LiChrospher, Spherisorb ODS1, Zorbax ODS or Nucleosil 100.
- **State-of-the-art polar phases:**
  e.g., "embedded phases" such as XTerra RP, Nucleosil Nautilus or Prontosil ACE, or even more strongly polar phases such as Fluofix, SynergiPOLAR RP or Platinum EPS.
- **Phases with a short chain and/or a polar functional group:**
  e.g., C_1, CN or diol.
- **Phases with a particularly small (50–60 Å) or a particularly large pore diameter (300 Å):**
  e.g., Nucleosil 50, NovaPak, Symmetry 300 or Jupiter.
- **"Outlandish" but promising experiments** (to be tried initially on a clean column, which can otherwise be discarded):
  e.g., use an ordinary C_{18}/C_8 column and as an eluent just water acidified using perchloric acid (pH value around 2 to 3) or a C_8 column and 30% acetic acid, or simply pump an Ag^+ or Cu^{2+} (AgCl or $CuSO_4$) solution over a C_{18}/C_8 or a silica gel column using a 40–60% water/acetonitrile eluent. You could also try using a polar column (e.g., silica gel) and separate hydrophobic isomers in pure methanol or acetonitrile. If you have to work with a buffer use a 0.5 M phosphate buffer and a middle-of-the-road (with respect to hydrophobicity) column, e.g., Spherisorb ODS 2.

I would also like to mention the following phases that are renowned for their selectivity in isomer separations, although I have no practical experience with them:

Callixarene, polymer gels, Hypercarb, C_{30} with hydrophilic endcapping.

Choice of eluents

Considering the organic portion of the eluent, my first choice would be methanol. The peaks may be broader, but its selectivity is higher compared with acetonitrile. I would also work with unbuffered mixtures rather than with buffered ones.

Conclusion

When separating isomers you have to rely primarily on ionic/polar interactions, not just the normal interaction with polar groups on the RP surface, but also try to exploit the ion exchange capacity of the SiOH groups. Alternatively, you could also look at the steric aspect and use a stationary phase with a small/large pore diameter (see Tip No. 3).

Tip No. 06

When should I use a "polar" C_{18} phase?

Problem/Question

In recent years, a number of hydrophobic C_{18} and C_8 phases have been introduced that are low in metal ions and have good coverage. The strength of these phases lies mainly in the excellent peak symmetry achieved when analysing basic compounds. Moreover, if used to determine other organic molecules they have proved highly selective, even when polar groups are involved. In the light of all these advantages, should such phases be given preference when developing a new method?

Solution/Answer

Not necessarily, because the separation of some types of analytes rely heavily on polar interaction. Phases with distinctly hydrophobic, well-covered surfaces cannot be expected to perform well under these circumstances (see also Tip No. 4).

For a variety of reasons, columns that have a high polar functionality have a better selectivity for the following types of analytes:

- Large hydrophobic aromatic compounds
- Positional and double bond isomers
- Strong acids (in dissociated form)
- Strongly polar metabolites
- Planar/non-planar molecules

Some examples of phases with polar functionality:

- Non-endcapped phases, e.g., Spherisorb ODS 1, LiChrospher
- Phases with polar groups on the surface, e.g., Platinum EPS, Supelcosil ABZ PLUS
- Phases with a polar group embedded in the alkyl chain, known as embedded phases, EPG, e.g., Nucleosil Nautilus, Hypersil ADVANCE, SymmetryShield RP_8
- Hydrophilic endcapped phases (only to a certain extent!), e.g., YMC AQ, Reprosil AQ
- Combination of the above, e.g., short alkyl chain plus an embedded polar group, such as SynergiPOLAR RP
- Specialized phases, e.g., with steric and chemical protection or a short, fluorinated chain, e.g., Zorbax Bonus, Fluofix INW

Conclusion

What is true for nature and everyday life also applies to HPLC, the more specialized a species (column), the better adapted it is to perform a certain task (separation). The results speak for themselves. There is always a trade-off between the high performance of a specialist and the less brilliant result from a jack-of-all-trades.

Tip No. 07 — Are polar RP-C_{18} phases more suitable for the separation of polar analytes than non-polar phases?

Problem/Question

We have recently seen the introduction of a large number of polar RP phases, most of which can be classified into two main groups: the polar endcapped phases ("AQ", "AQUA") on the one hand and phases containing a polar group embedded in their alkyl chain on the other, also known as polar-embedded phases. Their polar character is defined by a typically shorter alkyl chain (C_8, C_{12}, C_{16}) and, of course, a polar group, usually carbamate, amide or urea. Can we conclude from this recent development that such phases should be our first choice when separating ionic analytes?

Solution/Answer

Yes and no! It depends on the ionization state of the analyte or, from an even more general consideration, on whether the polar character of the analyte is prevalent or not. Let us first look at some of my recent results, from a selection of separations carried out in an acidic or neutral medium.

1. 3-Hydroxy- and 4-hydroxybenzoic acid, acidic compounds not dissociated at the acidic pH value (around 2.7) that we used, behave like neutral molecules in an RP system and therefore separate better on a hydrophobic Discovery C_{18} phase than on the more polar Discovery C_{16} Amide.
2. Phthalic and terephthalic acids are still in their ionic forms in this acidic medium. Since they are ionic and thus polar analytes, they separate better with the Discovery C_{16} Amide than with the more hydrophobic Discovery C_{18}.
3. Phenol/caffeine separate better in a neutral medium using SymmetryShield RP instead of XTerra RP. Both columns feature a polar carbamate group at the surface. SymmetryShield, however, is the more polar of the two because its matrix is silica gel, whereas XTerra uses a hybrid material with CH_3-groups on the surface and in the matrix, which gives it a more hydrophobic character.
4. Owing to their distinct organic character, tricyclic antidepressants can be separated in an acidic medium. The organic character is apparently so dominant that XTerra MS (hydrophobic surface, good coverage) achieves a better separation than XTerra RP. XTerra RP with its carbamate group is less suitable for the separation of weak organic bases.

Conclusion

The predominant character of the analyte (ionic/neutral, depending among other things on the pH value of the eluent) determines the choice of column. At a given pH value, if the analytes are found in an undissociated form or, due to an organic residue, they are non-polar and hydrophobic in character, a non-polar C_{18} phase is recommended. For an ionic analyte, polar RP (C_{18}) phases are probably preferable, if there is sufficient selectivity.

Tip No. 08 What about non-endcapped phases – are they a thing of the past?

Problem/Question

In the last years, a number of new C_{18} phases have been introduced, such as chemically protected phases or polar-embedded phases, polar endcapped phases, hybrid phases and monolithic phases. These materials have many advantages, but does this mean that we should always use one of the state-of-the-art phases when developing a new method?

Solution/Answer

No. In the separation of two similarly hydrophobic substances that differ in the arrangement of substituents or feature double bonds in a side chain (α,β-isomerism, positional isomerism), residual silanol groups have a decisive impact on the selectivity of the phase (see Tip No. 05), as demonstrated in Figure 8-1.

The top and centre chromatograms show the injection of uracil and three steroids (α,β-isomers) on two modern hydrophobic phases. Steroids Nos. 2 and 3 co-elute. The chromatogram at the bottom of Figure 8-1 shows the successful separation on Resolve C_{18}, which is an older packing material that is not endcapped. Such non-endcapped phases are also an excellent choice when acidic compounds have to be separated without a buffer. The resulting peaks generally have a better shape.

Conclusion

While for many separations state-of-the-art endcapped materials are definitely the right choice, there are cases (e.g., positional isomers and strongly acidic substances) where the selectivity of non-endcapped phases is higher due to the residual silanol groups

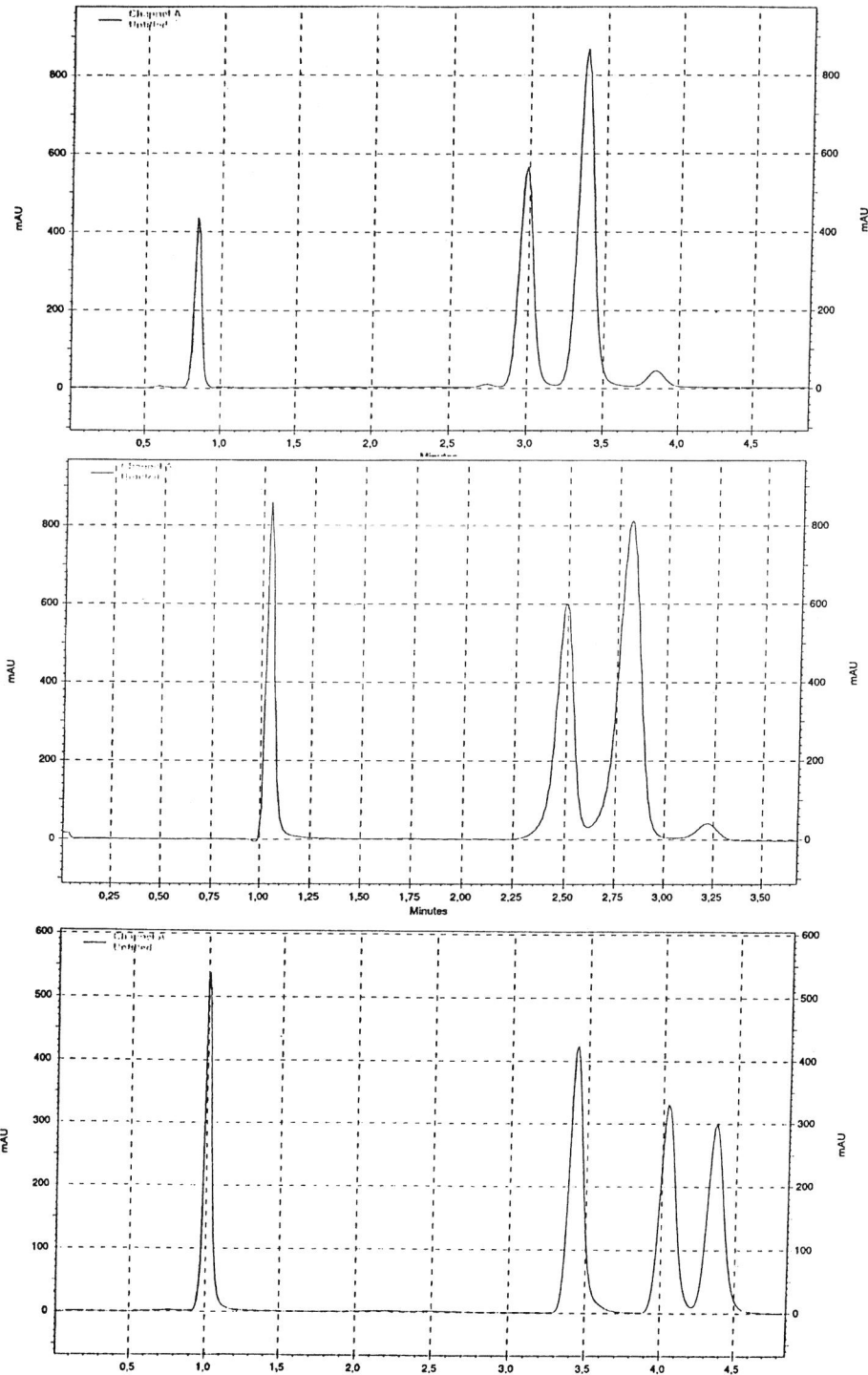

Tip No. 09 How can I separate acids using RP C_{18}?

Problem/Question

Suppose you want to separate acidic organic compounds. Let us also assume that these are not amino acids – there are excellent standard applications available for these, and to be honest, MS-MS coupling is so much more rapid and elegant as a method that it will soon replace the HPLC analysis of amino acids altogether. Your substances don't fall into the category of strong organic or inorganic acids either, as you would then use strong or weak ion exchangers to separate them. So which RP systems should you consider?

Solution/Answer

Use the following set-up:

1. The column
   "Classical", non-endcapped columns because of their greater stability in an acidic medium, such as LiChrospher, or some of the newer columns developed especially for use in an acidic medium such as Zorbax SB.
2. The eluent
   The eluent has to be acidic, because only in an acidic medium can the acidic components remain undissociated, and only then can they interact with the C_{18} surface, which, in turn, is a precondition for a reasonable retention.
   Weak acids should be separated at a pH of about 4–5 (eluent containing acetic, formic acids), stronger acids at a pH of about 2–3 (eluent containing phosphoric, acetic acid generally produces an improved peak form in comparison with phosphoric, trifluoroacetic acid). For a pH of about 1.5–2, perchloric acid has proved to be an excellent choice, but also consider methane sulfonic acid. There are other interesting, more exotic alternatives, which I will not mention here.

   Should we just use the acid or do we need to prepare a buffer?
   As a general rule, using buffers (in this case acetate, phosphate, formate, trifluoracetate and perchlorate in the form of potassium, ammonium or sodium salts) contributes to the robustness of the result. More on this in Tip No. 15. Bear in mind, though, that the critical range is above pH 4.

3. Other factors
   Further steps towards optimization:
   - Adding ion pair reagents, such as tetrabutyl ammonium chloride or tetrabutyl hydrogen sulfate
   - Lowering the temperature

Figure 8-1. Separation of three steroids using two endcapped phases (top and centre) and a non-endcapped C_{18} phase (see text for further explanations).

- Including a short polar column, such as CN in a serial arrangement with C_{18} – you will be amazed how neatly polar components can be separated (see Tip No. 34).

Conclusion

To put it simply, separating acids by HPLC resembles GC separation. It either works or it does not, and apart from the measures just mentioned, there is not much one can do to optimize the separation.

Tip No. 10 — The nitrile phase – some like it polar

Problem/Question

Many HPLC users seem to think that RP phases other than the usual C_{18} or C_8 phases are something exotic. Indeed they are. For example, the nitrile phase, also known as the cyano phase is one such unusual polar phase. When should it be used? What type of rules can we derive for its use?

Solution/Answer

A nitrile phase has a polar as well as a non-polar side to it. The CN group accounts for its polar character, while the propyl group (the link between the CN group and the silica gel surface) gives it non-polar properties. Thus, with respect to polarity, CN sits in the middle between C_{18} and silica gel.

This means:
If non-polar components interact too strongly with C_{18} or C_8 (resulting in tediously long retention times), you could think of using a CN phase. If, for example you want to separate polar substances from neutral components it is an ideal choice. There is hardly any retention as far as the non-polar, hydrophobic compounds are concerned. They elute at the dead time or shortly after, while the polar components follow later.

Another example:
If your analytes take a long time to elute, you could of course add more acetonitrile to the eluent or perform a gradient separation. However, a cheaper and certainly more environmentally friendly alternative would be to use a CN column, but keep in mind that the life span of a CN phase is normally shorter than that of an alkyl phase (C_8, C_{18}).

As a phase of medium polarity, CN can also be used in normal phase mode in connection with hexane, heptane etc. as the mobile phase. It is not as polar as silica. Here is a list of its characteristics and possible advantages over silica gel:

- Traces of water left in the eluent (which results in a layer of water on the surface of the stationary phase) are less noticeable. Results are therefore more reproducible.
- Adding polar substances to the CN eluent is a quick way of modifying selectivity, and it works faster than with silica gel.
- Using silica in a gradient separation is often touch-and-go: using a CN phase makes it so much easier!
- In general, CN is more robust than silica gel.

There is, however, a major drawback to using the CN-phase – there is a risk of the CN group being irreversibly hydrolysed into hydroxylamine or it can form carboxylic acid.

One more thing:

It is of course possible to use a CN column with hexane one day, then give it a rinse and use it for a separation that involves an RP eluent, and *vice versa*. In other words, you could use the column sometimes in normal and sometimes in RP mode. At least, that is the theory. In practice, it is safer to work with one CN column in normal phase mode (the manufacturer will send it with the correct solvent) and reserve another one for use in RP mode. Some manufacturers supply CN phases specifically for normal-phase and for reversed-phase applications.

Conclusion

- CN is more polar than the usual alkyl phases C_{18}, C_8 and C_4. It is suitable for the separation of highly polar substances though with a considerably lower stability in comparison with the alkyl phases.
- CN is less polar than silica gel, and separations using CN are more robust than those using silica gel.

Should you want even higher polarity, think diol. Here you could use, for example, heptane plus one drop of water, this system is more stable than the other example of hexane and silica gel.

Tip No. 11

The selectivity of RP columns

Problem/Question

As we all know, the separation mechanism that determines selectivity in RP chromatography can be a complicated issue at times. Are there any rules that govern this process? Yes, there are, and some of them are given below.

Solution/Answer

From a wide range of experiments with various substances and eluents, the following rules about the suitability of polar and non-polar RP phases can be deduced.

Suitability of polar RP phases:

1. Unbuffered eluents
   - Hydrophobic, unsubstituted "large" aromatic compounds
   - Planar/non-planar aromatic compounds
   - Isomers (positional and *cis/trans* isomers)
2. Buffered eluents
   - Basic substances, however good selectivity is often overshadowed by a messy appearance of the peaks (peak tailing)
   - Moderately strong acids (in dissociated form)
   - Strongly polar metabolites

Suitability of hydrophobic RP phases:

1. Unbuffered eluents
   - Polar and non-polar small neutral organic molecules (aldehydes, hydroxy benzoates, mononuclear aromatics)
   - Analytes that differ in polarity, the difference in polar character may be due either to a group ($C=O$, CH_2, etc.) or the result of isomerism
2. Buffered eluents
   - Weak acids (in undissociated form)
   - Organic bases

Which RP phases can be considered as "polar"?
The polar character of C_{18}-alkyl phases decreases in the following order:

- Polar groups on the surface, e.g., Supelcosil ABZ PLUS, Platinum EPS
- High overall concentration of silanol groups, e.g., Resolve, Spherisorb ODS1
- Embedded phases (C_{18} to C_3)
- Endcapped but silanophilic, metal ion-containing phases, e.g., Bondapak
- AQ-phases

I would also like to mention combination phases (e.g., SynergiPOLAR RP), specialized phases (e.g., Fluofix) or classical polar RP phases such as CN or diol.

Conclusion

- Non-polar substances can be separated selectively in an unbuffered eluent on polar phases. With increasing polarity of the analyte, there is a greater need for non-polar phases.
- Polar substances can be separated in unbuffered eluents on non-polar phases, yielding good results. This is also true for substances that differ in polarity due to the presence of an isomeric form or a diverging substituent, e.g., OH or H vs. CH_3 or perhaps CH_2 vs. C=O.
- Polar substances can be separated in buffered eluents on polar phases, while non-polar analytes require non-polar phases for good separation.

In order to achieve good selectivity, unbuffered eluents call for opposites in analyte and phase polarity, while buffered eluents call for similarity.

Hints

1. These rules apply to a substantial number of the analytes examined. However, to be valid in general, they would have to be verified for additional classes of substances.
2. Suitability here only refers to selectivity, not to peak symmetry.

1.2 Buffers, pH Value

Tip No. 12 — **Does it always have to be potassium phosphate?**

Problem/Question

Potassium phosphate seems to be by far the most popular buffer in RP chromatography. Potassium dihydrogen is used for the acidic pH range, while dipotassium hydrogen phosphate covers the neutral and weakly alkaline range of the spectrum. First of all, I would like to emphasize that if you are happy with your potassium phosphate, by all means stick to your guns, continue as before and go on to the next page. If, however, you are not particularly happy with it, think about ammonium phosphate, which has several advantages over potassium phosphate, as listed below.

Solution/Answer

- Ammonium phosphate is available in greater purity, thus reducing the likelihood of interfering peaks in sensitive gradient separations.
- Ammonium phosphate has a better solubility, which reduces the risk of unwanted precipitates when using 80/90% acetonitrile.
- Filtration is not always necessary.
- Ammonium phosphate has a lower UV absorption, which reduces drift in the lower ranges of the UV wavelength spectrum.

Conclusion

Ammonium phosphate may offer a viable alternative for the determination of trace substances, possibly in combination with high acetonitrile concentrations and low wavelengths.

Tip No. 13

UV cut-off of buffer solutions

Problem/Question

In trace analysis, the interference of an eluent with strong UV absorption is undesirable, particularly in the lower wavelength range. Acetonitrile rather than methanol would be the obvious choice of solvent, and, of course, it must be of "gradient" quality. Unfortunately, however, any additives to the eluent (modifiers, buffers, ion-pair reagents, etc.) only increase its intrinsic UV absorption. Thus, for example, after adding 1% acetate to acetonitrile, the level at which sensitive measurements can be taken rises to above 250 nm. What about the cut-off point (the wavelength below which no measurement can be taken, as the intrinsic absorption by the eluent becomes too strong) of commonly used buffers?

Solution/Answer

In the list given in Table 13-1, you will find the UV cut-off values for a number of buffers and their working pH range. These values were found in literature, and thus have not been measured by us. (Source: John W. Dolan, BASi Northwest Laboratory).

If the ionic strength is below 20 mM, the UV absorption is of course lower.

Conclusion

Adding various additives to the eluent usually improves the shape of the peaks and sometimes even the selectivity. However, these advantages could be outweighed by a decrease in the signal-to-noise ratios. If this is the case, choose a low ionic strength, or, treating selectivity as is given, you may decide to focus on the detection limit, which could mean leaving additives out altogether. The art is to find a viable compromise between robustness of the result and a sensible limit of detection.

Table 13-1.

pH value	Buffer/modifier	UV cut-off (nm)[a]
2.0–3.0	phosphate	210
2.5–7.5	citrate	250
3.5–6.0	acetate	230–240
6.0–8.5	phosphate	210
7.0–9.5	"TRIS" (tris-hydroxymethylaminomethane)	220–225
8.0–10.5	borate	210
9.0–12.0	diethylamine (fresh!)	210

a) These values apply to about 50 mM of buffer.

Tip No. 14 — Sources of errors when using buffers

Problem/Question

When separating ionic analytes, buffered eluents are indispensable, as they render the results more reproducible. However, a variety of problems may crop up, such as a lack of stability in retention times, changes in peak shapes or a short lifetime of the column. What could be the causes?

Solution/Answer

Please keep the following possible causes in mind:

1. Wrong buffer in relation to the desired pH range, e.g., phosphate buffer for pH=5.
2. Buffer too weak, e.g., separation of strong bases using a 5 mM buffer: 10 mM or 20 mM would be a safer bet.
3. Chromatography within ±1 pH unit around the pK_a value of the analyte, e.g., separation of aniline at pH 5.5.
4. The standard operating procedure (SOP) does not tell you whether the buffer salt is in the hydrated form or not. This would have an impact on the ionic strength, e.g., sodium hydrogen phosphate with 1, 2, ... water molecules.
5. There is a noticeable difference between the pH levels of the eluent and the silica gel base of the packing (see also point 6), e.g., a weakly alkaline buffer (pH value around 7.5) and LiChrospher (pH value of the base silica about 3.5).
6. It could be forgotten that following the addition of acetonitrile or methanol the pH value of acidic buffers drifts towards the alkaline. If you use a phosphate buffer with a pH of 7.6 and add 70% methanol you will end up with a pH of about 8.4. In this context you should also remember unintentional pH gradients that result from separations where solvent A and solvent B differ in pH value.

The most common cause for the shift in retention time in buffered systems is a change in pH. Clearly other causes such as a malfunctioning pump must be excluded. Checking whether your system is stable with respect to the pH value is pretty straightforward.

Simply measure the following three pH values:

1. pH level of the ready-to-use eluent, i.e., after adding acetonitrile or methanol. Obviously, owing to the proportion of organic solvent this is not the same pH value as measured in water, but here we are talking of comparative measurements only.
2. pH level of eluent after 2, 8, 24 h, depending on for how long you run your separations.
3. pH level of the eluate leaving the column.

Ideally, you should end up with three identical pH values.

If you find major inconsistencies, your system is flawed, and you should check whether one of the causes given above applies to your system.

Conclusion

Getting a system to work with ionic analytes and buffers may be troublesome, but think how much more trouble it would be to use ionic analytes with an unbuffered eluent in routine separations! There is nothing for it but to find a reasonably robust buffer. When running isocratic separations, use the recycling mode and an Elusaver to discard the peak to waste. With these tools, you can establish a well-balanced and reproducible method.

Tip No. 15 — The drawbacks of using buffers

Problem/Question

When separating polar/ionic components as a matter routine, buffers are indispensable. What are the drawbacks when using them?

Solution/Answer

Here is a list of possibilities:

- The intrinsic UV absorption of the eluent increases, pushing up the detection limit and perhaps disturbing the baseline.
- The lifetime of the column is reduced because an increase in ionic strength also increases the polarity of the eluent. Following the principle of like "dissolves like", the highly polar eluent now has the ability to dissolve the, also polar, silica gel.
- While the polarity of the eluent increases, the retention time of polar components decreases (they prefer a polar environment). This may lead to premature elution and perhaps insufficient resolution.
- Very important – buffers tend to even out the selectivity between stationary phases. The individual characters of the various phases are usually reflected in their polarity. This is partly lost through the use of buffers, particularly if acetonitrile is used as an organic solvent. Figure 15-1 shows the injection of uracil as a marker and 4-, 3-, 2-nitroanilines in an alkaline acetonitrile/buffer on four very different columns, see Table 15-1.

The chromatograms look very similar.

By contrast, if only methanol/water is used as an eluent, the different phases come into their own. Let us go back to the two embedded phase separations, now with a methanol/water eluent, see Figure 15-2. The elution order is actually reversed.

Conclusion

As mentioned previously, the advantage of using buffers lies in the stability of the separations obtained. The presence of buffers, however, may even out variations in

Table 15-1.

Symmetry shield:	C_8 and C_{18} "embedded phase" with carbamate as polar group
Zorbax bonus:	C_{14} "embedded phase" with amide as polar group and two diisopropyl groups as steric protection groups
XTerra MS:	Hybrid matrix, classical, hydrophobic C_{18} phase
Nucleosil HD:	Classical silica gel, classical hydrophobic C_{18} phase, extensive coverage

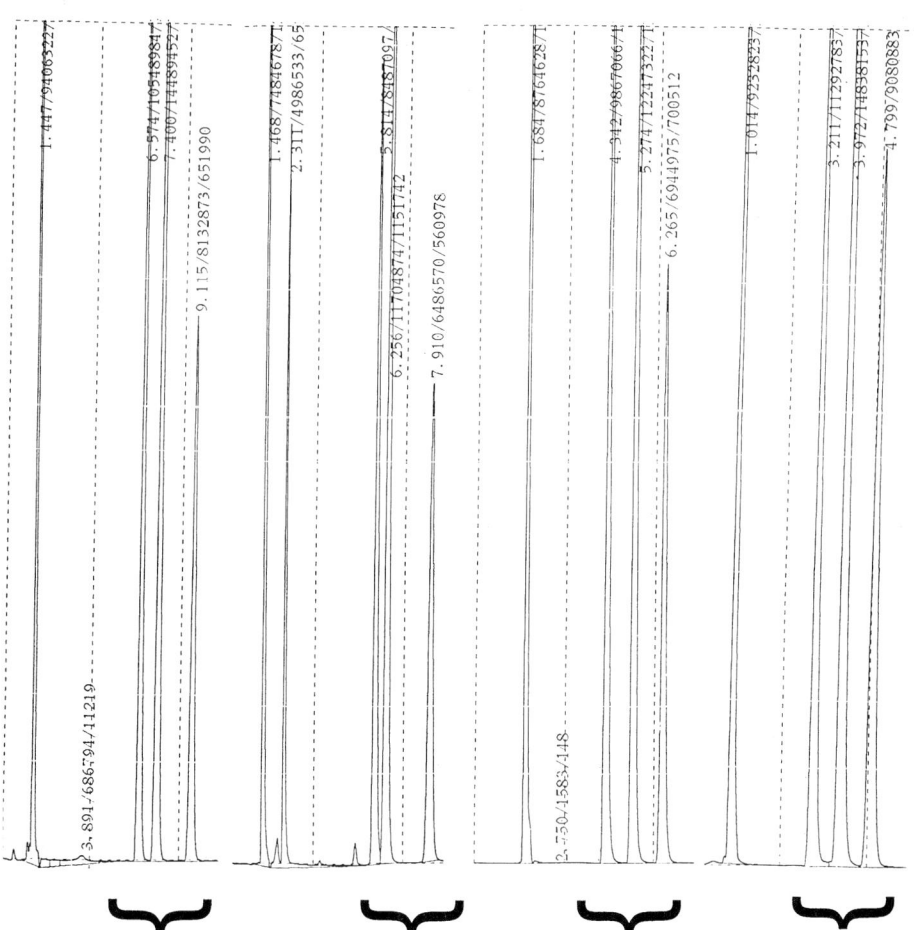

Figure 15-1. Separation of positional isomers in an acetonitrile/phosphate buffer on four different types of columns. Comment see text.

the polar properties of the phases. The phases appear to become similar to each other, the stationary phase can hardly be used as an optimization tool. The more polar the analytes to be separated, the more pronounced the lack in selectivity will be, as for example in highly polar metabolites. The disadvantage of using more selective non-buffered systems obviously lies in the difficulty of reproducing retention time and peak shapes. While for intricate one-off separations this may be a price worth paying, for routine separations the use of buffers is a must.

MeOH/H$_2$O (40/60)

Symmetry Shield

Zorbax Bonus

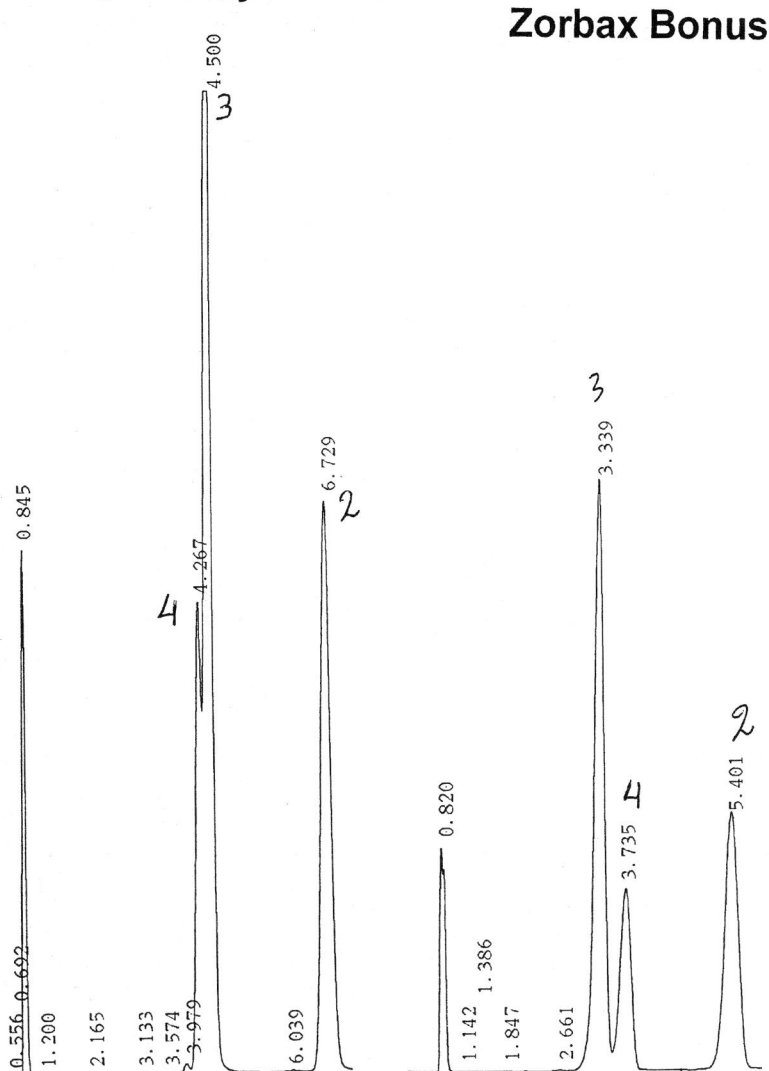

Figure 15-2.

Tip No. 16 — Why is the pH value so important, and what does it do?

Problem/Question

Anyone who works with polar/ionic or ionisable analytes knows that the pH value of the eluent is crucial: but what exactly does it do?

Comment 1:
This is such a complex matter that one could easily spend a whole day in a seminar or write a 10-page article on the subject. Here, we will only outline a few basic points.

Comment 2:
We will concentrate on RP-HPLC, as things are pretty clear-cut in ion exchange and ion chromatography: there is usually only one mechanism involved, the ion exchange mechanism.

Solution/Answer

The pH level affects the degree of dissociation of acidic/basic compounds and of free (residual) silanol groups on the surface of the stationary phase. This means that the pH value could be used as a means of controlling the degree of interaction between an polar/ionic analyte and the stationary phase. Thus, pH can be used to influence retention time and peak shape. Let us first concentrate on what happens with the residual silanol groups. We have the following equilibrium:

$$SiOH \iff SiO^- + H^+$$

Only if a silanol group carries a negative charge, can it interact strongly with basic compounds. This slows down the kinetic desorption process of the basic analytes from the stationary phase, resulting in pronounced tailing or even irreversible sorption. By using an acidic eluent, we can shift the equilibrium to the left (law of mass action!), which leaves us with undissociated silanol groups. They have been tamed now and will not attack our analytes, and we will be rewarded by a show of neat symmetrical peaks. This is why many SOPs say "add phosphoric acid" or "add phosphate buffer" or perhaps acetic acid or trifluoroacetic acid, etc. The acidic eluent ensures that peaks elute symmetrically, as the silanol groups are deactivated and therefore unable to interfere. However, if we want to separate stronger bases we have to work in a neutral or even alkaline medium, as this is where the bases, in their neutral form, can interact best with a non-polar RP surface. In an acidic medium, stronger bases carry a positive charge, and being ionic, polar species they would simply whiz through the column. They would produce symmetrical peaks all right, but within such a short time that we would stand no chance of separating them. There is nothing for it but to work with these bases in a neutral or alkaline medium.

So far, so good, but there is a small problem. From a pH value of around 4–5 the now negatively charged silanol groups are waiting for their chance to pounce on the bases, they just won't let go, and often ugly tails (the dreaded "chemical tailing" in bases) are the result. If we want our bases to interact with the C_{18} alkyl chains in peace we must distract the silanol groups by offering them a decoy: another base. This is why many SOPs say "add triethylamine", "add diethylamine" or some other base to the eluent or simply use an alkaline buffer. By using such additives with an eluent it is even possible to obtain reasonable peaks from older, non-endcapped materials such as LiChrospher. Whether a robust method is possible on these phases and systems is a completely different question.

Conclusion

Remember:

- An acidic eluent is the right choice for the separation of acids (good selectivity and good peak symmetry) and weak organic bases (good peak symmetry, but selectivity?). The organic character of the latter may be sufficient to trigger interaction with the non-polar stationary phase and thus result in sufficient resolution. It is therefore a good idea to begin your methods development and separation experiments using an acidic eluent.
- Stronger bases must be separated in a neutral/alkaline medium, there is no doubt about it, and endcapped phases are the first choice. The suitability of an additive for the eluent or the suitable pH range of the buffer is determined by the ionization of the analyte, i.e., by the basicity of the compound to be analysed. See also Tip No. 4.

Tip No. 17 — Why does the pH value shift even though I am using the correct buffer and the buffer capacity is sufficient?

Problem/Question

You are working with an RP system and using the following eluent: methanol (or acetonitrile)/potassium dihydrogen phosphate buffer, pH 7.5, 45/55 (v/v) to separate basic substances. Your separation looks too good to be true. In order to save additional time, you increase the methanol concentration to 80% or run a 40 → 80% MeOH gradient, and all other conditions are unchanged. Now let us see what is happening!

The bases do not elute as early as they should. Instead, some or even all of them have made a giant jump towards longer retention times, and/or their selectivity changes. It could also be the case that the column does not last as long as it should. What is causing all this trouble?

Solution/Answer

Increasing the concentration of methanol or acetonitrile will shift the pH value of the commonly used acidic buffers towards the alkaline. Depending on the ionic strength of the buffer, a pH value of between 8.3 and 8.5 would be measured. Up to an ACN or MeOH proportion of about 20–30%, the shift is not very pronounced. What effect does this shift in pH value have?

1. Stronger interaction of basic analytes with the stationary phase leads to an increased retention time.
2. Increased activity of residual silanol groups, increased peak asymmetry with pronounced and sometimes unacceptable tails.
3. The pK_b or even the pK_a value (pH level at which equal numbers of dissociated and undissociated molecules are found) shifts, and, depending on the actual pH level of the eluent, the selectivity of the separation may also change.
4. In a UV detector, a change in the peak area may be the result, as UV absorption could depend on the ionization of the analyte and thus on the pH value.
5. The increased alkalinity of the eluent shortens the lifetime of most silica gels, which explains why columns that are run at a nominal pH level of 7.0–7.5 do not last very long.

Conclusion

When preparing a buffer (measuring the pH value of the aqueous component of the mobile phase before adding ACN/MeOH) and during gradient separations one should keep in mind that the resulting pH shift needs to be taken into account, or the effects mentioned above could be the consequence. Thus, at least checking the pH value of the eluate that leaves the column is good practice: see also Tip No. 14.

Tip No. 18

Changes to the pH value in the eluent: the extent of the shift and the reasons behind it

Problem/Question

The longer I study the role of the pH value in the RP-HPLC separation of acidic and basic compounds, the more it strikes me how complex a subject this is and what an impact it can have. Here I just want to concentrate on one single aspect: the reasons for an unintentional shift. Under what circumstances can we expect a pH shift?

Solution/Answer

Table 18-1 gives the results from some of my recent experiments.

Table 18-1.

Reason for the shift	Extent of the pH value shift
After adding MeOH or ACN the pH value shifts towards the alkaline (see Tip No. 17)	Examples: a) pH of H_2O = 5.2, after adding 40% MeOH or 32% ACN pH value 7.4 ... 7.6 b) potassium phosphate buffer set at pH 2.68, after adding 32% ACN new pH-value 3.15 c) potassium phosphate buffer set at pH 7.59, after adding 40% MeOH new pH value 8.51 d) MeOH/phosphate buffer [40/60 (v/v)], 20 mM, original pH value 7.40. After about 400 injections of basic samples, the pH value of the eluent increased to 7.98
During the chromatography of basic compounds and isocratic separations in recycling mode without Elusaver, a more or less noticeable shift of the pH value can be observed	Shift in the region of 0.3–0.8 pH units
With increased ionic strength, the pH value also slightly shifts towards the alkaline, whereas the choice of the cation hardly seems to affect the pH at all 20 mM LiH_2PO_4: pH 3.16 20 mM $NH_4H_2PO_4$: pH 3.20	5 mmol KH_2PO_4/MeOH (60/40) pH 2.96 20 mMol KH_2PO_4/MeOH (60/40) pH 3.15 40 mMol KH_2PO_4/MeOH (60/40) pH 3.24
If the buffer does not match the pH level, there may be a shift. Read more about this in Vol. 1 Tip No. 27	

Moreover the salt concentration and therefore the buffer capacity after the addition of methanol/acetonitrile is reduced due to the dilution.

What are the consequences?

An uncontrolled shift in the pH value can compromise the robustness of a method, as retention times as well as selectivity or peak shapes can be affected. See Figure 18-1.

For example, adding MeOH raises the pH level of a phosphate buffer from an initial pH 3 to about pH 4. If you anticipate a shift on a similar scale, it makes more sense to use an acetate buffer instead. In the bottom panel of Figure 18-1 (acetate buffer) Ibuprofen yields a symmetrical peak. Its retention time is shorter and remains constant. Another factor that may have an impact on the results is the pH value of the original material [1].

An extra hint: special glass electrodes are recommended when using organic solvents to measure pH levels!

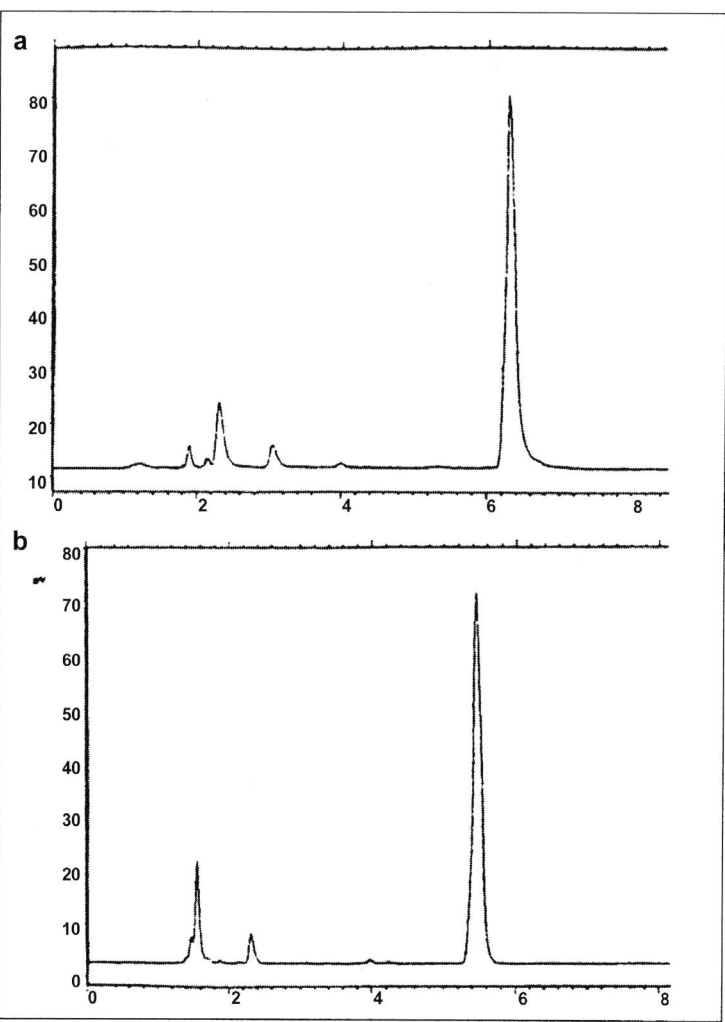

Figure 18-1. Injection of Ibuprofen at a pH 4 with 60% ACN and 40%: a, 5 mM phosphate buffer and b, 5 mM acetate buffer.

Conclusion

- When working with buffers in the "neutral" pH range, remember that the column might not last very long because the actual pH value of the eluent drifts towards the alkaline, where many columns are unstable: see Tip No. 17.
- Reproducibility can become a problem in the weakly acidic range.
- In order to achieve robust separation, the best choice is either a highly acidic or a highly alkaline medium. This means that we only have to deal with one type of molecule, either 100% protonated or 100% non-protonated, and we are well away from the critical range around the pK_a value of the acid/base. It is in this range where both types of molecules occur that tailing has been observed. One should, however, use columns that are made specifically for extreme pH levels.

Recommendation:

Obtain the $pK_{a/b}$ values for the known components in your samples. It is worthwhile, when needed, to use software tools (e.g., ACD) to calculate these! This is because this is an interesting area both for selectivity, and also for robust experiments.

Tip No. 19 An unintentional pH shift and its consequences

Problem/Question

An unintentional pH shift in the eluent is one of the main reasons for a lack of robustness in RP chromatography. In Tip No. 18 we were discussing the causes and the extent of such a pH shift. Now we are looking in more detail at the changes in peak size.

Solution/Answer

A pH shift can lead to a variation in signal intensity (peak area), as the UV absorption of ionisable compounds can be affected by the pH value. Figure 19-1 shows the spectrum of L-ascorbic acid at different pH values. When we look at the pH range relevant to HPLC, namely pH 2 to 9, we can see, for example, a smaller peak at 240 nm in the alkaline. At pH 5 it would be considerably higher, somewhat smaller at pH 4, and in an acidic medium (pH 2) it would reach its highest level. Now let us look at Figure 19-2, the pH dependency of the UV spectrum of three barbiturates. Considering the absorption spectra at 240 nm for pH 6.0 and 7.7: the difference in signal intensity is enormous for barbital, small for thiogenal and minimal for phenprocoumon.

Conclusion

The peak area can change as a result of an intentional (method optimization) or an unintentional pH shift. This may affect the individual peaks within the same chromatogram in different ways. In this context, I just want to remind you that the pH value also depends on the temperature. However, now let us look on the bright side – the pH value can also be used as a tool to improve the detection limit in trace analysis.

Here are some of the effects a pH shift can have:

- A change in interaction results in a change in retention time. Keep the possibility of a reversal in the elution order in mind!
- A change in the kinetics, resulting in tailing.
- Prototropic equilibrium of interconvertable isomers: one substance may appear as two peaks. This may change with pH. Example: proline-containing molecules such as captopril.
- Chemical change in the original component, yielding new substances and several peaks. In these cases, one should also remember that the stationary phase could also have a catalytic effect. For example, silica gel is popular as a solid catalyst in organic synthesis.
- Suppose a basic eluent contains methanol. This makes the eluent even more basic. As a result, the silica gel also dissolves more quickly and thus reduces the lifetime of the column.

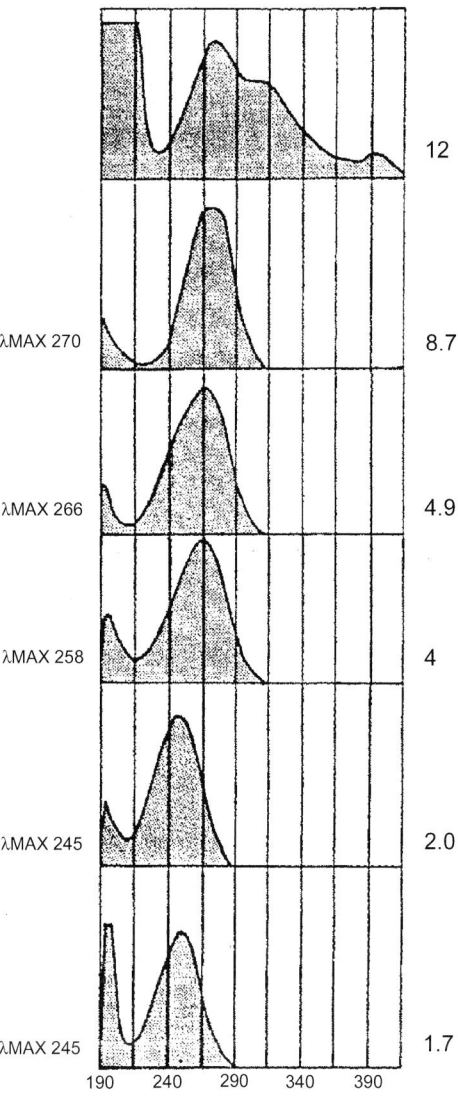

Figure 19-1. pH dependence of the UV absorption from L-ascorbic acid.

One last hint:

If you make coffee at 8:30 a.m. and put it in a thermos flask it will still be quite drinkable at 11:00 a.m., whereas coffee prepared at the same time but kept on the warming plate of a coffee machine is an acquired taste, because of a whole host of carbonic acids that develop – quite a challenge for the tastebuds and the stomach. From coffee back to sample preparation: if the sample solution is left to stand for some time, even at a normal temperature, the pH value and with it the chromatogram may change, so please check the consistency of your pH!

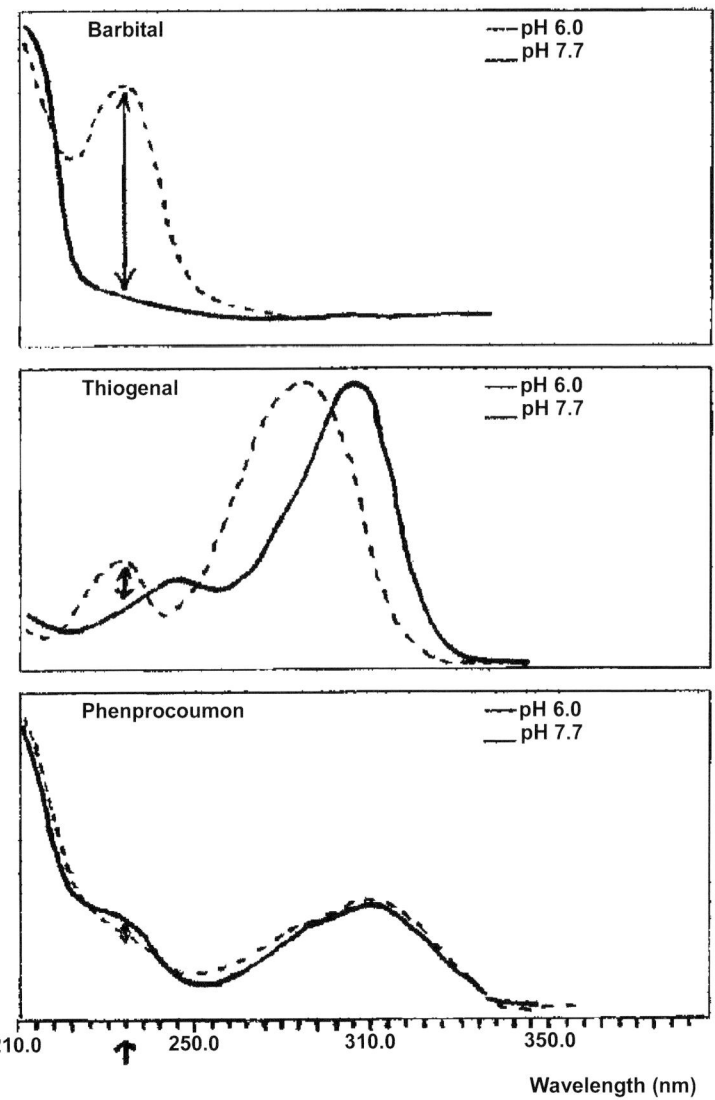

Figure 19-2. pH dependence of the UV absorption spectra for three barbiturates. Comments see text.

Tip No. 20 — RP separations in the alkaline medium

Problem/Question

An RP separation at a pH value of, say, between 7.5 and 9.5 is no fun. Suppose that for selectivity reasons an alkaline pH value and a column based on silica gel instead of polymer gels is required. Depending on the column type, this may have a greater or lesser impact on the longevity of the column, and the robustness of the results may also be affected. What is to be done?

Solution/Answer

If you are tied to an SOP, then just do what you are told, there is no room for discussion. If, however, you are allowed some freedom, then you could consider the following measures (the simplest ones are marked with an x).

- Column
    - Strongly hydrophobic phases with a high surface coverage above 2.5 µmol m^{-2} are fairly stable in an alkaline medium.
    - Hybrid phases, e.g., the XTerra-series.
    - Phases with bridged bonding, e.g., Zorbax Extend.
    - A phase with a polymer layer, e.g., Gromsil CP or SMT OD C_{18}.

- Eluent
    - Use organic buffers instead of phosphate buffers. TRIS or borate buffers fit the bill.
    - If you have to use phosphate buffers, make sure it is Li, Na or K. Avoid NH_4-salts!
    - Eluents containing ACN are usually gentler to the column than those containing MeOH.
    - The salt concentrations (ionic strength) in the eluent should be kept to a minimum.

- Other measures
    - x • Use a conditioning column (saturation column) between the pump and injector: see Tip No. 07/1.
    - x • Keep the temperature below 40 °C wherever possible.
    - x • Even if only a minimal decrease of the pH level can be achieved this will have a positive effect not only on the lifetime of the column but also on the reproducibility of retention times. Find out whether the separation could perhaps also be done at pH 7.8 instead of pH 8.3!
    - x • Finally, flush the column with MeOH/H_2O or ACN/H_2O, weakly acidic (pH ≈ 3–4) for storage.

Conclusion

Just one or two of the measures mentioned above could noticeably lengthen the lifetime of your column. However, it is usually not worth going to extreme lengths just to prolong the life of a column. After all, the price of a column makes up a small percentage of the entire cost of an analysis, in the region of 1–2%; but then again, if it helps to achieve a robust method, it is certainly worth the trouble.

Tip No. 21 Separation of basic and acidic compounds contained in the same sample

Problem/Question

A variety of state-of-the art columns are now available that are perfectly suited to the separation of strongly basic analytes. By contrast, when it comes to the separation of strong acids, returning to non-endcapped classics is usually advisable. As a rule of thumb we can say:

1. For strongly basic analytes (i.e., in protonated form) RP phases with good coverage and low in silanol activity should be used combined with an alkaline eluent. See also Tip No. 4.
2. For weaker organic bases that have a pronounced organic character (in the undissociated form) use phases as in 1 combined with an acidic eluent.
3. For weaker acids (undissociated) use phase and eluent as in 2.
4. For stronger acids (dissociated) use silanophilic polar RP phases combined with a (strongly) acidic eluent.

What if a sample contains basic as well as acidic compounds?

Solution/Answer

In this case, the answer is less clear-cut. As I mentioned before, non-endcapped silanophilic materials can be useless for strong bases, resulting in pronounced peak tailing or even irreversible adsorption onto the stationary phase. Conversely, materials with good coverage tend to yield strongly tailing peaks if very acidic components are injected; see Figure 21-1, injection of the highly acidic phthalic acid on two state-of-the-art well-covered C_{18}-materials. The few still remaining silanol groups form hydrogen bridges with acids. This in itself would not be worrying, the only snag is that due to the scarcity of silanol groups in these well-covered state-of-the-art materials, the kinetics are slowed down (cause: overloading). Thus, tailing becomes inevitable.

In this case one should either resort to endcapped classics, which still contain a considerable number of silanol groups, or use modern phases of medium polarity. The dissociation status of the silanol groups and the analytes is crucial and can be manipulated by choosing the appropriate pH value of the eluent.

Take a buffer of low acidity, around pH 4, and change its pH value by 0.5 pH units up or down. Columns that have sufficient polarity as well as hydrophobic properties should be used, e.g., LiChrospher, Spherisorb ODS 2, Zorbax ODS, Prontosil ACE, Purospher, MP-Gel or AQUA.

Conclusion

Suum cuique – each to his own, and don't ignore each other's idiosyncrasies – is a good motto in real life as well as in HPLC ...

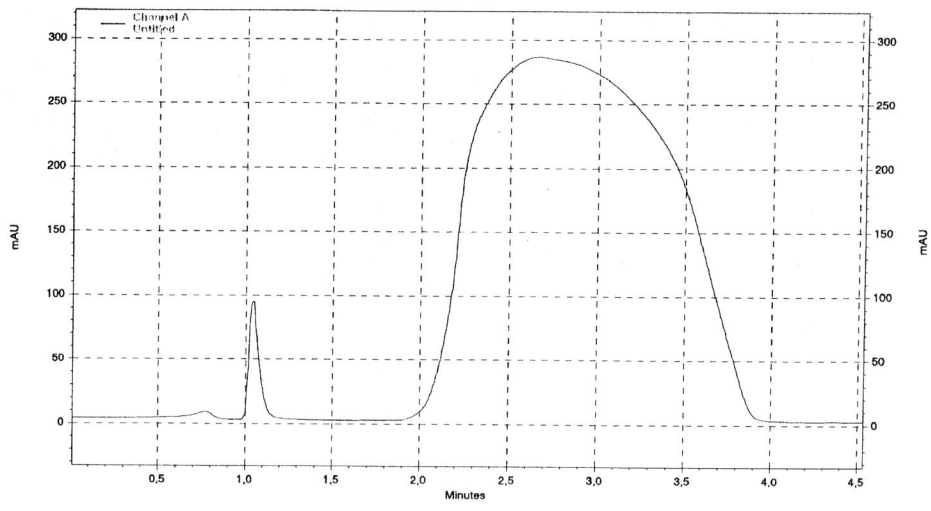

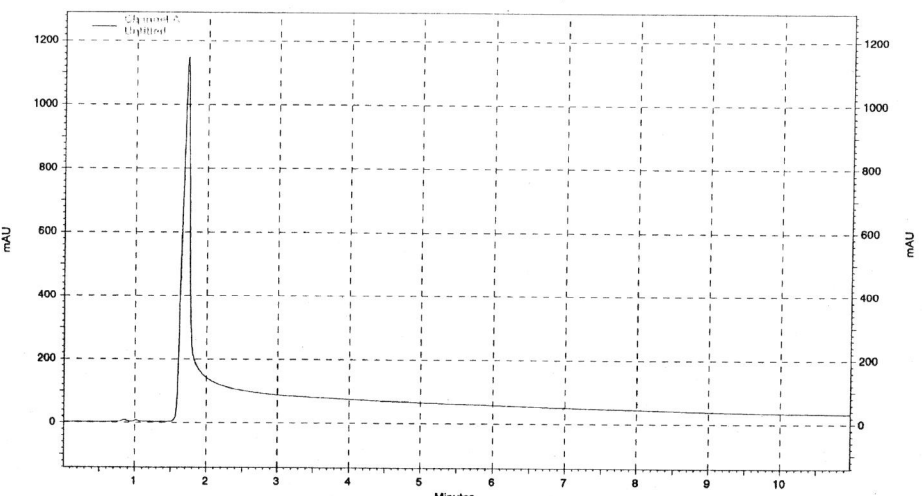

Figure 21-1. Peak shapes of strong acids on hydrophobic RP phases.

1.3 Optimization, Peak Homogeneity

Tip No. 22 — **The peaks appear too soon – what can be done?**

Problem/Question

Suppose that in an RP system (C_{18}, ACN/phosphate buffer 30/70) a peak appears too soon and you are not even sure if it is just one single peak – there could be more lurking behind it. It would appear that you are dealing with one or more highly polar compounds. What can you do to increase their retention time?

Solution/Answer

1. Modify the chromatographic parameters – use a longer column, slow down the flow and/or decrease the temperature. This can all be done fairly quickly, whatever the specific separation problem may be.
2. Modify the stationary phase:
   - Use a material with a smaller pore diameter, e.g., Si 60 Å instead of Si 120 Å, which increases the specific surface area and hence also interaction.
   - Use non-endcapped material.
     The (supposedly) ionic components can now interact with free silanol groups and thus elute later. The likely drawback would be some tailing.
   - Either use polar phases that have the ability to interact but are free of active silanol groups, which could cause annoying tails, or use phases with a large number of active silanol groups that speed up the kinetics. Suitable phases for polar analytes are, e.g., Zorbax SB C_8, Fluofix, SynergiPOLAR RP, perhaps also phenyl, nitrile or diol (see also Tip No. 6).
3. Modify the eluent – increase the water content, lower the buffer capacity, change the pH value (by far the most effective parameter!), add ion-pair reagents.
4. Combination of options 2 and 3.
   Be bold when choosing a stationary phase and an eluent for strongly ionic components. For example,
   - 30% acetic acid with a C_8 phase.
   - 100% H_2O with a diol phase.
   - Phosphate buffer + 100 mMol heptane sulfonic acid, pH around 2 with a non-endcapped C_{18}-phase.
   - Borate buffer, pH around 9 with an endcapped C_{18} phase with high coverage (more than 2.5 µMol m^{-2}) or a polymer-based stationary phase; see also Tip No. 5.

Conclusion

The most effective way of dealing with the problem at hand is through modifying the pH value, while changes in buffer strength, ion pair concentration and stationary phase only come second. If chromatography is to be carried out in a strongly alkaline medium on a silica gel-based column, state-of-the-art phases with high coverage, polymeric phases or hybrid materials are recommended because of their stability in this range, see Tip No. 20.

Tip No. 23 What can I do if the peaks elute late?

Problem/Question

Your separation yielded a good resolution – now your next question could or should be "What is the most trouble-free way of reducing the analysis time?"

Solution/Answer

Easy – simply increase the flow! I know, many of my readers think this a trivial answer. So it is, but I think it is still worth pointing it out in a world where far too many protocols just stipulate, "flow 1 mL min^{-1}". Increasing the flow always saves time and sometimes even improves the resolution. Apart from the smaller peak areas, the disadvantages are negligible.

Here is why –

Isocratic separation: Increasing the flow means saving time without noticeably reducing efficiency (theoretical plates), as today's spherical 5 or 3 µm particles have a very flat van Deemter C-term. The number of theoretical plates remains high, especially for real substances, i.e., if the peaks elute significantly later than an inert component. This has often been described in the literature as well as being confirmed by my own measurements.

Here are two simple examples.

Figure 23-1 shows the separation of four components on a 5 µm column at a flow rate of between 2 and 0.5 mL min^{-1}. Is the separation at 1 mL min^{-1} really significantly better than at 2 mL min^{-1}?

Figure 23-2 shows another separation at 2 mL min^{-1}, and the same separation in Figure 23-3 at a flow rate of 1 mL min^{-1}. Again, the improvement due to the reduced flow rate is only marginal.

If a major improvement is to be achieved, a really low flow rate must be chosen. Figure 23-4 shows the separation from the previous example at a flow rate of 0.2 mL min^{-1}. Peaks one and two are barely separated at 2 mL min^{-1}, and quite clearly at 1 mL min^{-1}, and at 0.2 mL min^{-1}, they are base-line separated. It depends very much on the individual case whether it is worth putting up with the excessive analysis time.

Gradient separation: Increasing the flow means saving time and/or an improved resolution.

At a constant gradient time t_{Gr}, an increased flow rate F results in an increased gradient volume, V_{Gr} ($V_{Gr} = F \times t_{Gr}$), and thus in an increase in peak capacity. As more peaks can be separated per unit time, resolution increases; or an increase in flow by a factor of 2 and a simultaneous reduction in gradient time, also by a factor of 2, yields the same separation at an analysis time which is reduced by a factor of 2.

Figure 23-5 shows a separation using a linear gradient at a flow rate of 1, 1.5 and 2 mL min^{-1} (flow increasing from top to bottom). While the separation takes 2.5 min

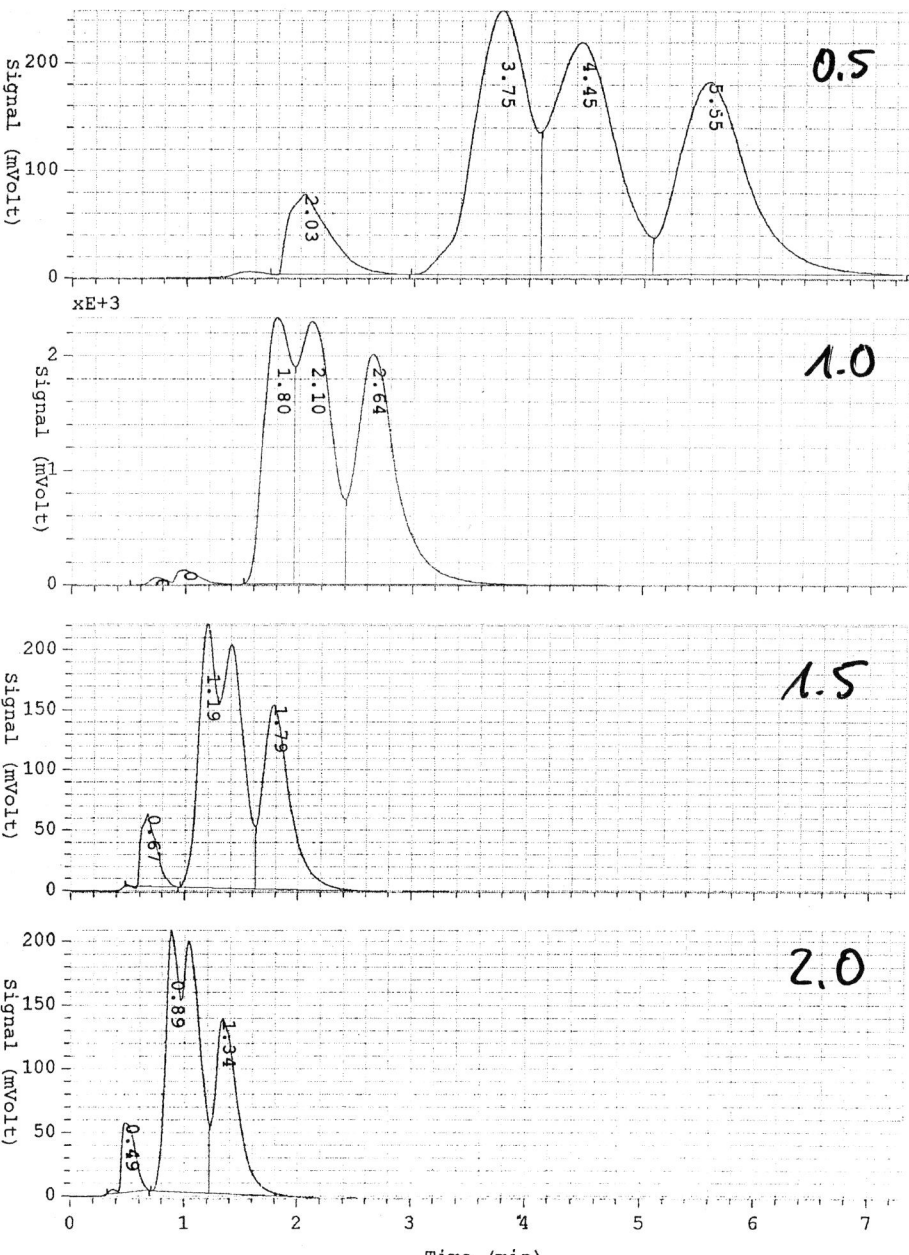

Figure 23-1. Separation of four compounds on a 5 μm column at flow rates of between 2 and 0.5 mL min^{-1}. Comments see text.

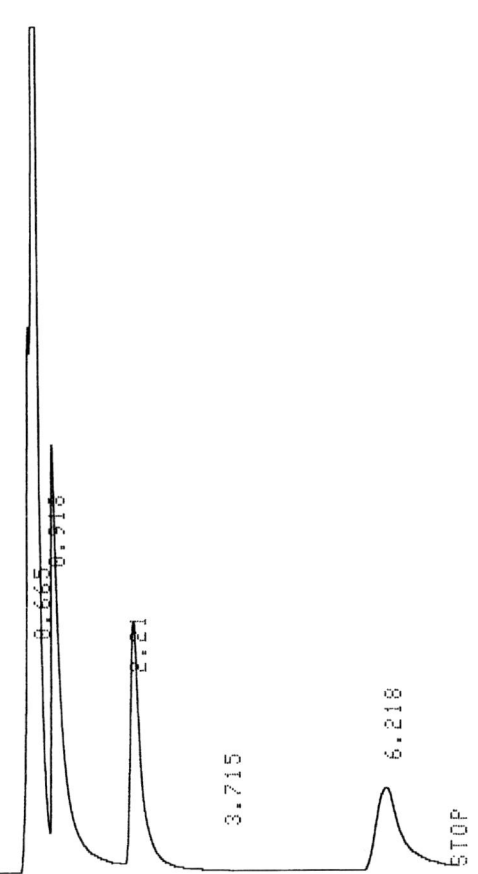

Figure 23-2. Separation of five compounds at 2 mL min^{-1}.

at 2 mL min^{-1}, it requires 4.6 min at 1 mL min^{-1}. At all flow rates, the resolution between peaks remains practically the same.

What about the drawbacks? If you are not dealing with trace analysis, I do not think they really matter.

1. Will the increased pressure shorten the lifetime of the column?
   The usual RP column withstands pressures of 300–350 bar without any ill effects. Just set the high-pressure limit on your pump to 4000 psi (1 bar = 145.038 psi).
   Besides, does it really matter if you use up four columns instead of three per year if you can save a significant amount of time?
2. Will the injector and pump seals wear out more quickly?
   Well, yes, you may have to spend a few cents more on them...
3. What if the column connection begins to leak when the pressure rises above 200 bar?
   In my opinion the time saving effect is such an overriding benefit that I would not hesitate to replace PEEK capillaries with steel ones.

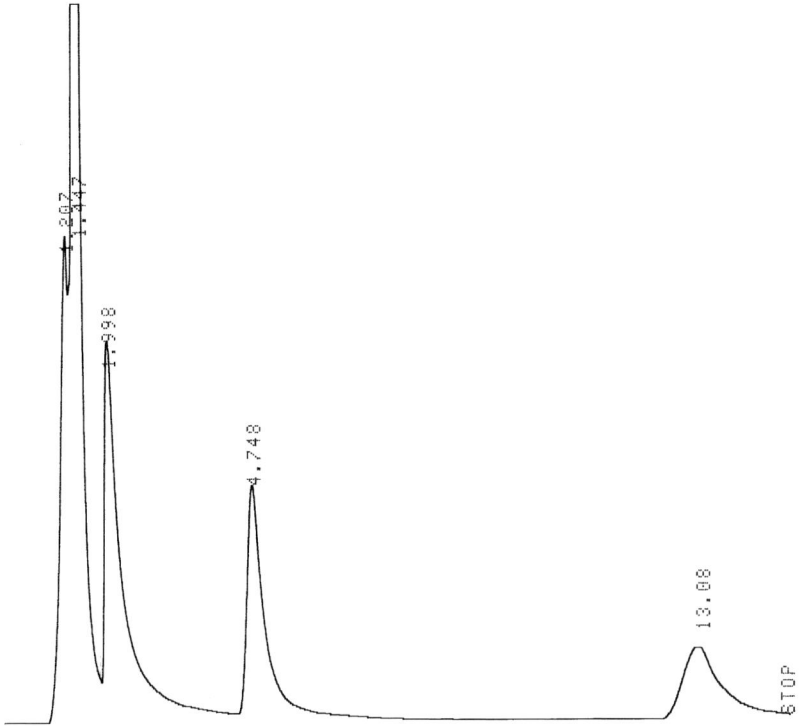

Figure 23-3. Separation of five compounds at 1 mL min^{-1}. Comments see text.

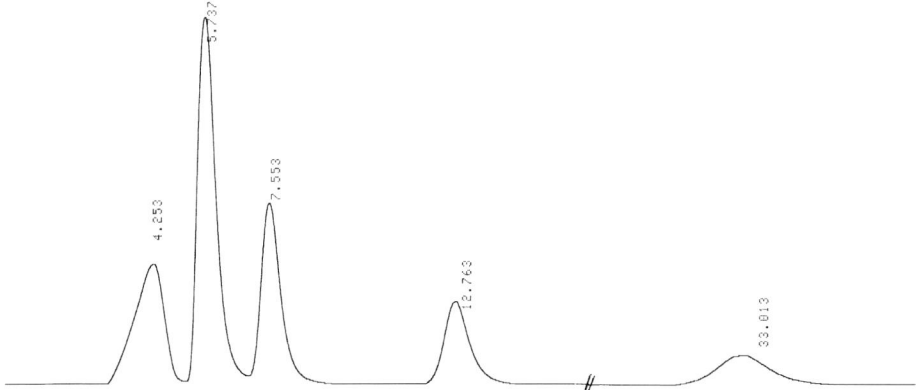

Figure 23-4. Separation of five compounds at 0.2 mL min^{-1}. Comments see text.

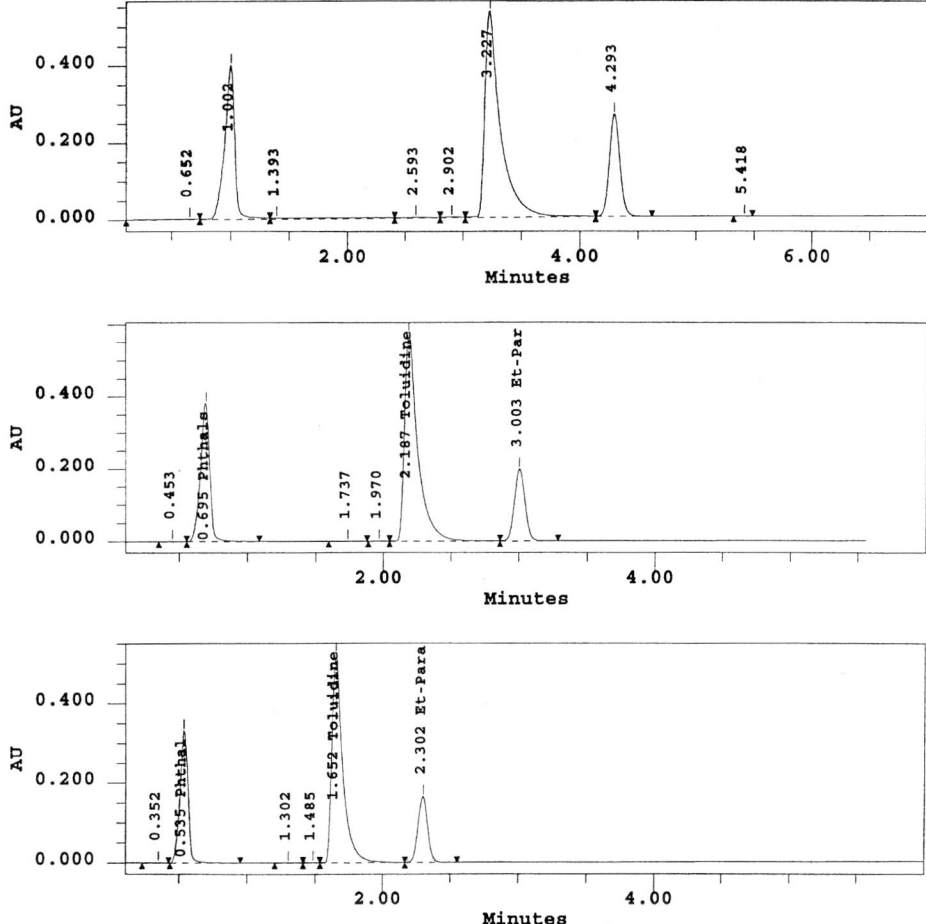

Figure 23-5. Reduction in analysis time, maintaining constant resolution, by increasing the flow rate with gradient separation. Comments see text.

Conclusion

In classical RP separations – i.e., those that do not involve ion exchangers, microcolumns or enantiomeric separations, exclusion chromatography, etc., flows of 1 mL min^{-1} are often far too tame. Unless, that is, you are struggling with peaks that are so small already that an increase in flow would diminish their area so considerably that you would have trouble detecting them.

One last word:

If you find the jump from 1 mL min^{-1} to 2 mL min^{-1} too drastic you could opt for 1.5 mL min^{-1}, which would still give you a time saving of about 30% – not bad, eh?

Tip No. 24 Quick optimization of an existing gradient method

Problem/Question

When I visited a firm that routinely ran many gradient processes, it turned out that not all of these methods worked satisfactorily from a user's point of view. We then tried to figure out how to improve one of their frequently used methods with minimum effort. This was successful, and I will describe for your benefit what we did:

Solution/Answer

Figure 24-1 shows the initial set-up, i.e., the usual methanol/water gradient at a flow rate of 1 mL min^{-1}. Thus, a chromatogram – not including regeneration time – took about 16 min, which we thought was far too long. We therefore increased the flow rate to 2.6 mL min^{-1} (pressure ca. 345 bar), reducing the retention time to about 10 min, a saving of approximately 30%, see Figure 24-2. However, we thought that this was still too long, so we set the gradient starting composition at 40% instead of 10%, Figure 24-3. Now the time was OK, but we did not like the shape of the peaks. We noticed from the front edge a hint that the solvent used to dissolve the sample was not optimal. It turned out that the guidelines recommended the use of tetrahydrofuran/acetonitrile – a fairly strong solvent mix in comparison with the eluent. We then

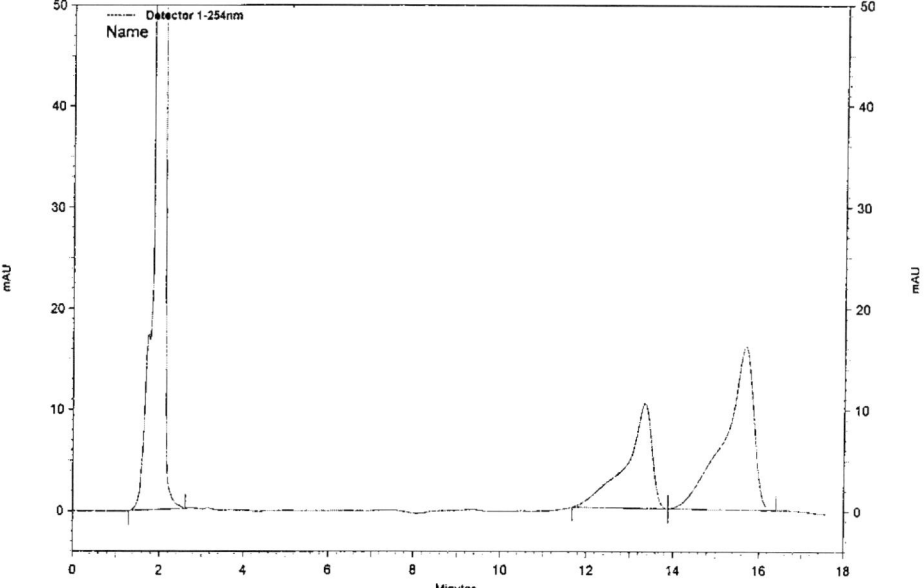

Figure 24-1.

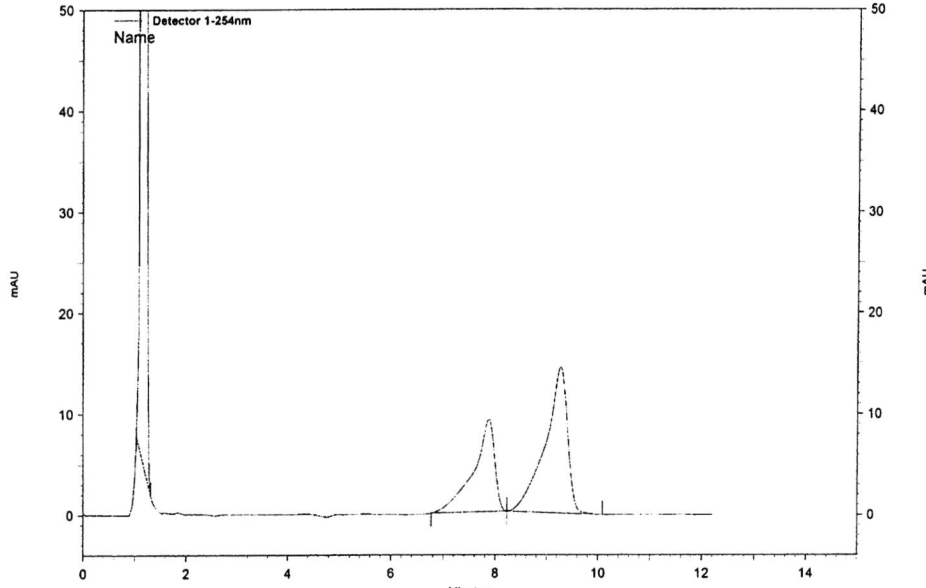

Figure 24-2.

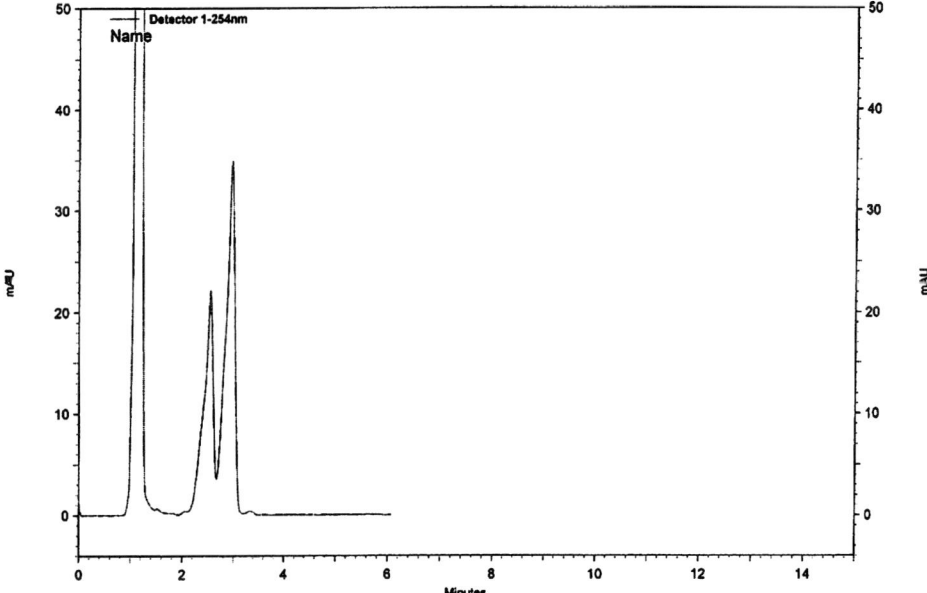

Figure 24-3.

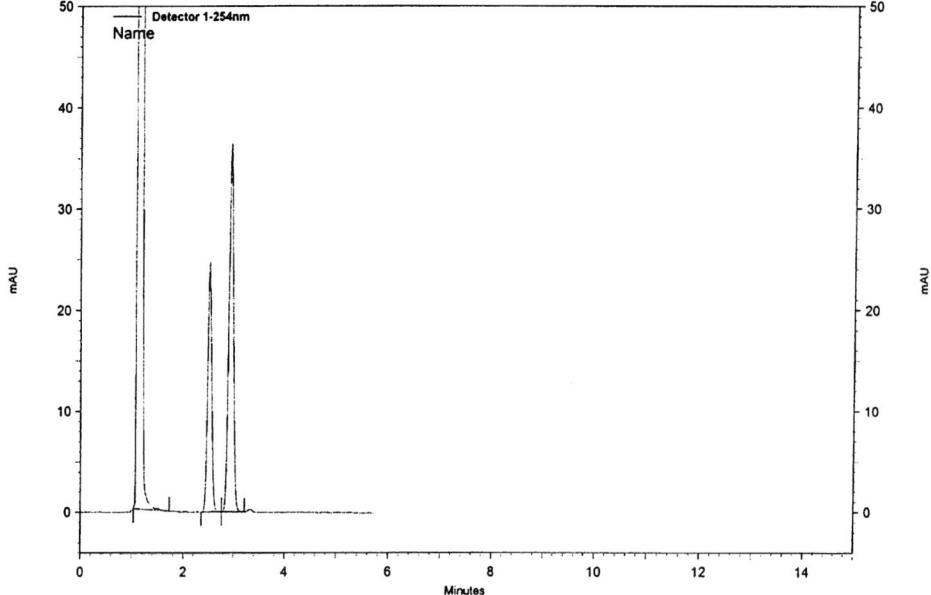

Figure 24-4.

diluted the sample solution with eluent A (40/60 methanol/water) twice and injected it again, Figure 24-4. Then we were happy with the result.

Conclusion

Here is a simplified description of a proven optimization strategy.

First of all, make sure that peaks are eluted within a reasonable time span, then – if selectivity is sufficient – think of rapidly improving the shape of the peaks. Only then is it worth running an other eluent or column (alternative method).

Applied to a gradient, this means:

1. Flow rates should be kept as high as possible – keep the gradient volume in mind!
2. You should start off with a high percentage of methanol/acetonitrile and use steep gradients.
3. The higher the polarity of the sample solvent compared with the eluent, the more likely it is that the analytes are concentrated at the top of the column, resulting in an immaculate peak shape. This aspect will become increasingly important with large injection volumes.

Tip No. 25

Increasing efficiency – often the fast track to success

Problem/Question

It should be stated at the onset that the most important and safest way of improving chromatographic resolution is increasing selectivity. The following examples should clarify this statement.

For a separation factor of 1.01, 160000 theoretical plates are needed to elute two baseline-separated peaks. If, with the help of pH value, modifier, etc., you can improve the separation factor to 1.05, only 6000 plates are needed, at 1.10, 1940 plates and at a separation factor of 1.20 no more than 575 plates.

When the separation factor is raised, e.g., from 1.05 to 1.10, then the improvement in the resolution is by a factor of 2. See also Chapter 3.2.

Suppose you have been experimenting with a given separation and reached some resolution that you still find is insufficient. Should you try to continue to increase selectivity, or are there other ways of achieving the desired resolution?

Solution/Answer

Suppose you achieved a separation factor a of 1.1, but you are still not happy with the resolution. This can only mean that your peaks are too broad (poor efficiency) and resolution needs to be improved. You have now two options to achieve this:

1. You try to improve selectivity further by modifying the chromatographic conditions, such as column, eluent, etc. As I said before, this can be rather tedious (improving resolution via selectivity).
2. You leave the chromatographic conditions and thus the selectivity unchanged but try to increase efficiency, i.e., the plate number. This often proves to be the quicker way (improving resolution via efficiency).

Let me give you two examples.

The left panel of Figure 25-1 shows the separation of tricyclic antidepressants on Luna C_{18}, 5 µm, in an acidic acetonitrile buffer. The a-value (separation factor) between peaks 3 and 2 is 1.05. In this case, improving selectivity is not a trivial task it takes time and some genius. On the right of Figure 25-1 you see a separation of the same material on a 3 µm column.

While selectivity remains nearly the same at $a=1.04$, baseline separation is almost achieved. The increase in pressure to 230 bar does not pose a significant problem.

The left panel of Figure 25-2 shows the separation of chrysene and perylene on Symmetry C_{18} in methanol. The a-value is 2.13, but the peak shape is nothing to write home about. If you replace the capillary that connects the injector with the column by a thinner one, you can reduce the dead volume within the instrument and achieve a somewhat better separation at an almost identical separation factor of 2.10. By the way, once I had made up my mind, the second improvement was done in 70 s!

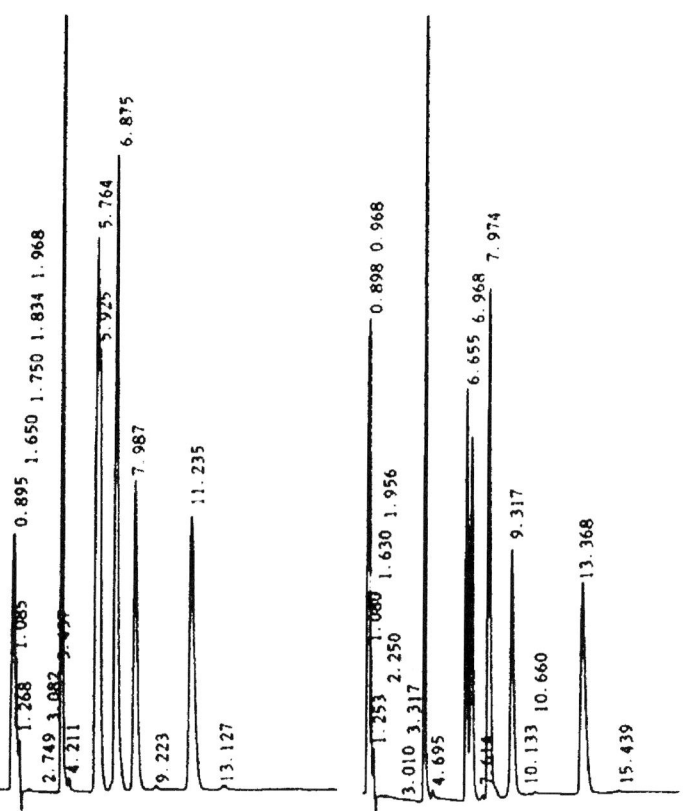

Figure 25-1. Influence of the particle size on resolution using the same conditions. Comments see text.

Conclusion

If the retention time and the selectivity are acceptable but the resolution is not, aim for maximum efficiency – in other words, make the peaks narrower – before you try out more laborious ways of optimizing your separation.

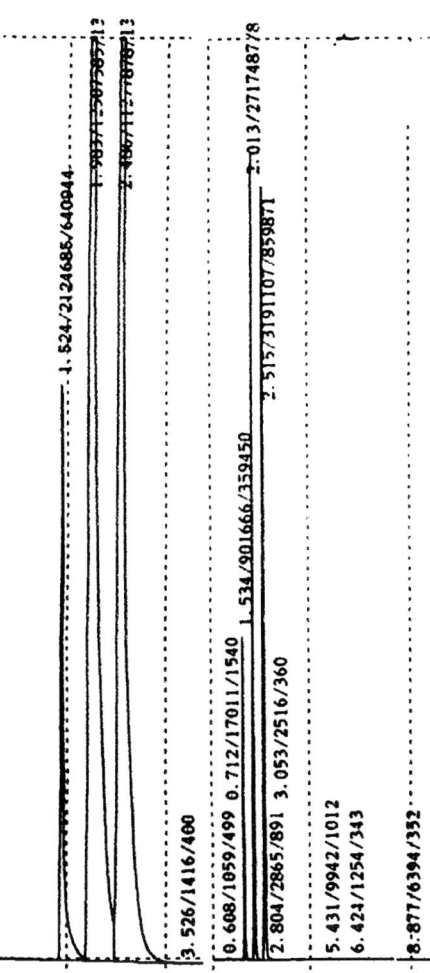

Figure 25-2. Influence of the dead volume on the resolution by isocratic separation. Comments see text.

Tip No. 26 — Additives to the eluent

Problem/Question

The preparation of eluents often involves small percentages of additives or modifiers that are mainly added to a binary eluent. Why is this necessary, and what is the effect of these additives?

Solution/Answer

In HPLC, the bag of tricks is practically inexhaustible. The main objectives are:

1. Improving peak shapes.
2. Modifying selectivity.

Table 26-1 lists the most common additives.

Conclusion

Small amounts of additives can have an enormous effect, but if you care for the robustness of your results, please bear in mind that the addition of additives could render some equilibria fairly labile. Keep the experimental conditions constant – constant temperature, constant pH value, constant concentration, consistent degree of purity of your reagents.

Table 26-1. Common additives to eluents.

When … and/or change	What	How much, comment	Comment
I. Peak shape/selectivity			
Separation of acids	Acetic acid/acetate	0.1–1%, pH around 4	• disadvantageous cut-off at 250 nm, aggressive towards steel
	Oxalic acid	0.1–1%, pH around 4	e.g., for tetracyclines
	Phosphoric acid	0.1–1%, pH around 3	e.g., for antioxidants
	2-Hydroxyisobutyric acid	0.1–1%, pH around 3	• advantageous cut-off at 210 nm
	perchloric acid	pH around 2	• interesting effects in a strongly acidic medium
Separation of stronger acids	PIC-A-reagents, e.g., tetrabutylammonium chloride or tetrabutylammonium phosphate or -hydroxide	50–100 mM, pH = around 7.5 For strong acids a lower pH value is needed to neutralize them	Sometimes negative peaks or ghost peaks

Table 26-1 (continued)

When … and/or change	What	How much, comment	Comment
Separation of bases	Amines, e.g., triethylamine – Octylamine – Diethylamine – Dimethyloctylamine	20–100 mM (=0.05–0.5%) 10–50 mM 10–50 mM 10–50 mM	Diethyl- and dimethyloctylamine seem to be most effective
Separation of stronger bases	PIC-B-reagents, e.g., tetra-, penta- etc. up to dodeca (lauryl) sulfonic acid	50–100 mM, pH = around 3.5	Hepta- and octasulfonic acid are the most popular, for strong bases use longer sulfonic acids (C_{12} to C_{16})
Separation of neutrals	TFA, isopropanol, n-butanol	ca. 5–10% of methanol or acetonitrile replace against one of these modifiers	Interesting selectivity effect

II. Retention time, selectivity

When … and/or change	What	How much, comment	Comment
Compounds with an O–CH_3 group	Acetonitrile	Around 5–10% added to an MeOH/H_2O eluent	Increase in retention time
Compounds with an O–CH_3 group	Tetrahydrofurane	Around 5–10% added to an MeOH/H_2O eluent	Decrease in retention time
Substances with a C–O–R group	Acetonitrile	Around 5–10% added to an MeOH/H_2O eluent	Decrease in retention time
Substances with a Cl group	Dichloromethane	Around 5% added to an MeOH/H_2O-eluent	Decrease in retention time
Substances with a phenyl group	Tetrahydrofurane	Around 5% added to an MeOH/H_2O-eluent	Decrease in retention time
N-Heterocycles, aromatic amines	N,N-dimethylformamide	Around 0.1%	Decrease in retention time (Source: V. Meyer, EMPA, St. Gallen, Switzerland)
Separation on silica gel, aluminium oxide	Water	10–50 ppm	A few ppm of water cause a major change in retention time and selectivity
Exclusion chromatographic separation on a diol phase	Salts	Small percentage	Avoiding sorption to stationary phase

Table 26-1 (continued)

When … and/or change	What	How much, comment	Comment
III. Other			
Separation of compounds that form complexes	Ethyldiaminetetraacetate (EDTA)	10–20 mM	EDTA forms complexes with metal ions (steel, stationary phase)
Preventing growth of fungi	Sodium acid	0.05–0.1%	NaN_3 elutes with the front in an RP system
In situ endcapping: blocks activity of silanol groups in RP columns and on glass surfaces	Trimethylchlorosilane (TMS)	Lower per mL range	Prevention of uncontrolled sorption of bases or large polar molecules (e.g., proteins)

Tip No. 27 Separating the unknown – where shall I begin?

Problem/Question

Let us suppose you have to get a separation done, and, of course, you are short of time. Needless to say that you know very little about the sample. The only colleague experienced in HPLC is on holiday, the lab supervisor is at an HPLC conference and will then travel on to his favourite spa. So, you are left to your own devices with a vial filled with an unknown substance and a note on your desk saying URGENT in huge letters. As a trustworthy member of staff you resist the urge to go home straightaway but decide to do your best – how should you go about it?

Solution/Answer

This example is, of course, purely fictional because in real life, some information is usually available. But just for argument's sake, let us take the fiction for real. Although we cannot develop a whole method in just a few lines, I can give you a few hints on how to proceed in order to obtain some initial peaks and what to do next to optimize the procedure.

Preliminary decisions: Is it a liquid sample? → Measure its pH value!
Neutral → ACN/H_2O or MeOH/H_2O as eluent, usual C_{18}-column, see below
Acidic → Eluent as above, pH=2...3, non-endcapped C_{18} column
Alkaline → Eluent as above, pH=7...8, endcapped state-of-the-art C_{18} column
Solid sample? → Dissolve
Soluble in H_2O, MeOH, ACN, isopropanol and mixtures of these? → C_{18} column
Soluble in hexane, ether, etc. → SiO_2 or another polar column

The chromatographic system (a description of the most frequent system, an RP system)

Sample solvent:	Dissolve sample in eluent or in a solvent as similar as possible to the eluent. Tip: A sample solvent containing a little more water than the eluent yields sharp peaks
Injection volume:	At first no more than 20 µL if possible
Column:	New-generation C_{18} is best, i.e., metal-ion free, base-deactivated, endcapped (for acids, see above!) 125×4 mm or 100×4 mm, 5 µm or 3.5 µm. Real experts tend to take a 50 mm column for some quick experiments
Flow:	1.5–2 mL min^{-1}
Temperature:	30–40 °C

Detector:	DAD (Diode Array Detector), if not available run the separation twice, at about 220 nm and 280 nm (LC-MS would of course be a great help. RI (Retractive Index) lets you see everything, but sometimes with a very small signal, and it takes half a century to get a stable baseline. Also, you can't run gradients with RI.)
Separation mode:	1. Gradient – the first step in the right direction

Run a $10 \rightarrow 90\%$ ACN, linear gradient at a rate of 2 mL min^{-1}, duration: 10–15 min

Variation: 1st run at pH ≈ 3 (phosphoric acid)
             2nd run at pH ≈ 7.5 (phosphate buffer with triethylamine)

Quick optimization options:

- Change gradient volume (flow or time, see Tip No. 23)
- Slope of the gradient
- Initial and final conditions
- Perhaps a variation in pH value

2. Isocratic (if you have no choice but to run an isocratic separation)

Start at about 60–80% ACN/H$_2$O and adapt the eluent strength in such a way that the first peaks elute at a retention factor between 2 and 5

Quick optimization options in order to recognize trends and check on robustness!

- ±5% ACN
- ±5 °C
- ±0.5 pH
- Tailing peaks? → Play around with the pH value
- Are they still tailing? Use ion-pair reagents and a well-covered phase
- Is there simply no end to the tails? Just put up with them, and when your boss returns from his spa, ask him extensively about his health. After all, the stay in the spa should have benefitted his health as well as his mood. Just use all your charm to convince him that all you need is a capillary electrophoresis instrument!

The strategy described should give you a first idea of the sample before you take further steps towards systematic optimization, such as changing the packing and the dimensions of the column, the strength of the buffer, adding a modifier, switching between ACN and MeOH, etc. This can be done with the help of commercial optimization programmes. Even if they do not make specific suggestions, if you know your way around them, they can still save you time.

Conclusion

It is very rare indeed to have as little information about a sample as in the situation described above. In any case, you should try and retrieve as much information as possible from the following sources: internal/external customers, internal/external literature databases, internet, manufacturer brochures, etc. They might just give you a first hint.

Tip No. 28 Separation of an unknown sample using a reversed-phase C_{18} column – how do I go about it?

Problem/Question

Suppose you want to separate a sample you know very little about, and you are using an RP C_{18} system. How do you choose the chromatographic conditions? This time, I want to give you a different point of view from that in Tip No. 27.

Solution/Answer

In the following, I will give you rough guidelines which may be adapted to individual circumstances.

1. Basic requirements

- The sample should be at least partly soluble in the eluent (H_2O/MeOH or H_2O/ACN).
- For isocratic separations the molecular weight should not exceed 600.
- For gradient separations the molecular weight can be much higher. If you do not know the molecular weight, you could use the following trick: elution before t_M (dead time) means that the sample is being excluded because the molecules in the sample are too large or their charge density is too high – see Tip No.73. Elution after t_M means that the molecules can diffuse into the pores.
- Where can the compounds be detected? Find the appropriate wavelength.
- Stability of the sample solution? Store a sample solution at room temperature for 24 h and in the refrigerator for 24 h. Then inject both samples and compare the chromatograms.
- Measure the pH in order to be able to choose a vaguely appropriate column for acidic/basic/neutral components.

2. The chromatographic set-up

- The sample solution is neutral: simply take a good C_{18} column (3–4 mm, 100–125 mm). Gradient: 10 to 90%, H_2O/ACN at 2 mL min^{-1} and 30–40 °C.
- The sample solution is acidic: take a good C_{18}-column (see above). Eluent: H_2O/ACN about 40/60 w/w, pH ≈ 3–3.5 (e.g., phosphoric, acetic or perchloric acid). For a very acidic solution, choose a polar, non-endcapped column. Eluent: H_2O/ACN plus tetrabutyl ammonium chloride or tetrabutyl hydrogen sulfate, ca. 2 g L^{-1} eluent.
- The sample solution is basic: state-of-the-art C_{18}-phase with excellent or even double endcapping, free of metal ions, protected phase, etc. Eluent: H_2O/ACN, see above, plus diethylamine or triethylamine, about 2 mL L^{-1} of eluent.
- For a strongly alkaline solution, eluent: H_2O/ACN plus about 50–70 mM hexane or heptane sulfonic acid.

In a second step, put the usual optimization procedures in place, such as varying the eluent, buffer and ionic strength, temperature, etc.

3. Validity of the result

If you have to be absolutely certain about the purity of the peaks (e.g., forensic toxicological evidence or high-risk agents) you should take a further step to validate your results.

(1) Use a diode array detector. This is a generally acceptable but often insufficient tool.
(2) Two subsequent columns plus diode array detection – a good method.
(3) Off-line/on-line chromatographic coupling: "Cut" the peaks and then proceed to separate the fractions over DC, GC and CE – a very good method.
(4) Proceed as in (3) and add off-line/on-line spectroscopy: MS, MALDI, NMR – a very, very good method.
(5) As in (3), but combined with an analysis based on a different principle, such as gel electrophoresis – an excellent method.
(6) As in (5) plus subsequent spectroscopy. This increases selectivity and specifity – the best and most expensive method of analysing unknown tricky samples.

Conclusion

If you follow the steps I have described you will have results very quickly, and which measures should be taken in order to guarantee the purity of the peaks depends on the importance of the sample.

Tip No. 29  Developing an RP separation – the two-day-method

Part 1: Choice of column and eluent

Hypothesis: Is it realistic to expect to have developed 80% of a method within two working days? Let us look at the following – fictional – example!

Problem/Question

Your boss comes into your lab on a Tuesday morning in a state of panic. He gives you an extremely important sample he got straight from the research lab. He implores you to do a separation before tomorrow night because he is going overseas on Thursday morning and urgently needs information about the number of components in the sample. So he wants you to pull qualitative analysis out of your hat, and as he is usually such a nice guy and not in the habit of making unreasonable requests, you are going to give it your best shot. How would you go about it?

Solution/Answer

We all know that there is more than one way to skin a cat. I will restrict myself to just one possibility.
You need at least:

- A gradient separation instrument with diode array detection (DAD)
- A 6-way column-switching valve or ideally a 12-way valve
- A collection of about 20 different RP columns

If you really don't know what is in the sample – which is rare in real life – or if you want to enhance your repertoire and flexibility – always a good thing to develop – the following requirements are recommended:

- An additional detector, such as a refractive index or a fluorescence detector or even LC-MS(MS)-coupling
- A 3- or 6-way eluent switching valve – unless you have a low pressure gradient
- A column oven with cooling capability

Procedure

Run a 10% → 90% linear H_2O/ACN gradient, e.g., at 2 mL min^{-1} and at a pH of around 3 (TFA, phosphoric or perchloric acid) over 20 min. There is nothing wrong with using your favourite type of C_{18} column, such as a 125 mm×4 mm, 3 or 5 µm, as long as it is a new one!

If you have the time, you should also consider the following variations.

1. Try also an H_2O/MeOH eluent!
2. Run one gradient at a pH value of around 7.5 and combine 1 and 2.

After the first run you will have a rough idea of how many peaks are in the sample. The gradient can then be slightly varied depending on the circumstances (e.g., initial and final ACN concentration, gradient slope, etc.) so that all peaks elute within a reasonable time span. It should take you 1.5 h to run an "overview" gradient. If you also try MeOH and/or a basic medium you would need about 3 h – everything running smoothly, that is. If you began your work at 8.30 a.m. you should have finished at 11.30 a.m. To be on the safe side, let us say 12.30 p.m.

Using your column-switching valve and timing the column switching using your software, run this roughly optimized eluent (or, if necessary, the corresponding isocratic mix) on the six columns linked to the valve. The time needed is about 30 min per column or 3 h overall. Meanwhile, you might as well go out for lunch, have a nice cup of coffee and get on with the rest of your work while the columns run on autopilot.

Now comes the crucial question – which six columns should you choose to run this first experiment?

Have a look at Tables 29-1 and 29-2. Table 29-1 lists phases that cover a variety of retention mechanisms in RP HPLC. Table 29-2 covers a wider range from hydrophobic to polar RP phases, listing some typical examples. The columns listed are examples of suitable types rather than recommendations.

Comments on the choice of columns:

Obviously, the composition of your column portfolio depends on the task at hand. Thus, the number of polar phases in Table 29-1 could be reduced in favour of additional hydrophobic phases. If you know, for example, that the sample consists of neutral organic molecules, you could leave out non-endcapped polar phases. Conversely, if you know that you have components of similar polarity but differing in structure, classical hydro-

Table 29-1. Various types of RP phases 1

Selectivity characteristic	Examples of columns	Characteristic feature of column
Steric aspect	Jupiter	Large pore diameter
	Nucleosil 50	Small pore diameter
Hydrophobic surface	Luna	Classical coverage
	Gromsil CP	Polysiloxane layer
Free silanol groups	Spherisorb ODS 1	Acidic silanol groups and low coverage
	Zorbax ODS	Acidic silanol groups and high coverage
Polar group in the alkyl chain (embedded phases)	HyPURITY Advance	Polar embedded phase
	Prontosil ACE	hydrophobic embedded phase
Strongly polar surface	SynergiPOLAR RP	Short alkyl chain, built-in polar group, polar end group, hydrophilic endcapping
	Platinum EPS	Polar groups at the surface
Free choice according to needs	e.g.,	
	Hypercarb	Very hydrophobic
	Fluofix IEW/INW	Very hydrophilic, fluorine atoms
	XTerra MS, etc.	Hybrid material, etc.

Table 29-2. Various types of RP phases 2

Luna (Inertsil ODS 3, Kromasil)	Purospher	Spherisorb ODS 2	HyPURITY Advance
SMT (Gromsil CP, Nucleosil HD)	MP-Gel	Platinum C_{18}	SynergiPOLAR RP
SynergiMAX RP (Zorbax Extend, XTerraMS)	Ultrasep ES	Nucleosil Nautilus	Spherisorb ODS 1
Nucleosil 50 (Nova-Pak)	Zorbax ODS	Symmetry Shield	Supelcosil ABZ PLUS
Jupiter	Reprosil AQ	Zorbax Bonus	Fluofix INW
Discovery C_{18} (YMC Pro, Hypersil BDS)	LiChrospher	Prontosil ACE	Platinum EPS
Hydrophobic phases plus steric aspects	"Polar" phases from the hydrophobic group	"Hydrophobic" phases from the polar group	Polar phases

phobic material will not be of much use. Take non-endcapped or just generally polar RP-C_{18} phases instead. If the sample consists of basic compounds, you might consider trying out three hydrophobic phases right from the start – two embedded phases and one polar, in order to check selectivity (see Tip No. 4) etc. The more you know about the sample, the more targeted your choice of columns can be. If, as in our fictitious case, you have no information at all, it would be best to try out one column of each type.

A 12-way valve could be connected with the following types of columns:

- endcapped/non-endcapped
- Si 60 Å/Si 300 Å
- Sterically or chemically protected/hydrophilic endcapped (SB, embedded, AQ)
- around 8% C/around 20% C
- Diol/phenyl
- Amine/nitrile

At about 3.30 p.m. just have a look at the six or 12 chromatograms to decide which column yields the best separation, i.e., where you got the largest number of peaks. Now connect another six columns to your valve that are similar to the best column in the first experiment. If, say, you notice that a non-endcapped, fairly polar phase shows the best selectivity, choose six polar columns for the second experiment.

Now run the same gradient over a second set of six columns overnight. The following morning, you just have to decide which of the 12 columns tested has the best selectivity. This will then be used for a fine-tuned optimization.

Variation:

If you can put an eluent-switching valve before your column, you could run this overnight experiment with several eluent compositions. The following morning, you could find out which combination of column and eluent has worked best. Now you have still about a day left to optimize your separation further.

Conclusion

A column-switching valve is an effective time-saving tool when it comes to choosing columns. Overnight experiments are highly recommended. You can, for example, find out in three overnight runs, which of 18 columns is best suited to deal with your specific separation problem – while you sleep!

Another possibility would be to run tests over four subsequent nights, each time testing six different columns of one type of phase. As these tests take place by night, their demand on staff working time is limited. Only the fine-tuning requires a member of staff to be actively involved.

Tip No. 30 Developing an RP separation – the two-day method

Part 2: Fine-tuning of the separation

Problem/Question

Following the procedure in Tip No. 29, you have found the optimum column, and you have around a day left for the fine-tuning of the separation. Where do you go from here?

Solution/Answer

Suppose you have many peaks, thus you will probably continue to work with a gradient.

1. Goal: Reasonable analysis time plus overall good separation
Options:
- Increase/decrease the gradient volume – best done by changing the flow rate. See Tips Nos. 23 and 24
- Vary the initial and final conditions
- Possibly change the slope and profile of the gradient or include an isocratic step

Time required: about 2 h.

2. Goal: Improve selectivity
Options:
- Vary pH value by ± 0.5 pH units
- Try out several organic modifiers, such as THF
- Lower temperature to 10–15 °C

Time required about 3 h.

3. Goal: Improve peak shapes
Options:
- Increase the polarity of the sample solvent. See Tip No. 24
- Check wavelength and other adjustable parameters and set them at their optimum. See Tips Nos. 31–34
- Inject a smaller amount of sample

Time required: about 2 h.

If your overnight experiment yielded 3 to 5 peaks that elute reasonably close to each other, you should perhaps now try to run an isocratic separation. In principle, what I mentioned in 2 and 3 is also valid for isocratic separations.

Other fairly quick optimization methods for an isocratic run:

- Use MeOH instead of ACN, perhaps also exchange about 5% of the organic phase with THF, which will have intriguing effects on selectivity. See Tip No. 26.
- Use a longer column or two consecutive columns – the ones that worked best in the overnight experiments. See Tip No. 34.

Of course, such optimization procedures can be carried out much more elegantly and rapidly if you have an optimization programme and you are reasonably familiar with it. However, even if your optimization skills are rather pedestrian, with a systematic approach you should be able to achieve good results within a day (approx. 7 h).

Conclusion

With a concept such as the one described in Tips Nos. 29 and 30, perhaps modified according to your needs, you can achieve a rough separation fairly quickly. If needed, further systematic fine-tuning can follow. Whether you need two or four days to develop your method – in any case, my advice would be:

- Be intelligently lazy – the night will do the work for you...
- This is the sequence to follow for the optimisation process:

1. Aim for a reasonable retention time.
2. When improving selectivity, try not to go overboard.
3. Improve the peak shapes, which also gives you an indirect double-check on peak homogeneity.

Tip No. 31 Quick check on peak homogeneity

Part 1

Problem/Question

You have finally optimized your separation, and your peaks look acceptable. Now the question is how to make a quick check on the homogeneity of the peaks. Quick in this case means that no major change is needed in hardware, e.g., adding a column-switching valve, or at a chemical level, such as using a different buffer. This tip and parts of Tips Nos. 32 and 33 give you simple measures that will only take a few minutes, 15 min at the most. Tip No. 34 will look at more labour-intensive methods.

Solution/Answer

1. Change of settings
   Of course, changing the settings on the instrument will not improve selectivity, but the peaks will become narrower. This will be particularly noticeable with early peaks. Some examples are given in Table 31-1.
   Changing the settings given in the table is a matter of seconds. Try it out, and you will be surprised how narrow your (early) peaks can become, and it is quite possible that you will discover one or two more small peaks as shoulders on others.
2. Small changes in the chromatographic set-up
   - **Reducing the flow rate**
     In 10 µm columns, which are quite often chiral columns, reducing the flow rate by 1.2 or 1 mL min^{-1} to 0.8 or 0.7 mL min^{-1} can significantly improve a separation. However, in 5 µm, not to mention 3 µm columns, one would have to reduce the flow quite drastically (as low as 0.3 or even 0.2 mL min^{-1}) to achieve a noticeable effect: see Tip No. 23.
   - **Concentrating the sample at the column inlet**
     Dissolve the sample in 70–80% water in a reversed-phase system. The remainder should be the organic component of the eluent. The peak shape improves.

Table 31-1.

Parameter	Modification
Response time, rise time, time constant, see Tip No. 68.	Decrease to 0.1 or even 0.05 s
Sampling time, sample rate	Increase to 4 to 5 data points per second
Peak width, see Tip No. 34/1	Decrease to 0.01 min
Slit width of the diode array	Large slit width lowers the detection limit
Detector output	Instead of using the 1 V output, use the 100 mV, or even better the 10 mV output

- **Reducing the injection volume**
  Column overloading, and peak broadening as a result of this, is more common than you might think. The risk is particularly high with ionic or ionisable components. Simply inject 5 or 2 µl instead of your normal 10 or 20 µl. If you get a better resolution, then you know what the problem was.
- **Change in wavelength**
  Just try your luck and inject the sample using two different wavelengths. If you have a diode array, this can be done much more elegantly.

Incidentally, these are all measures that could be carried out without too much fuss even when you are tied by tightly regulated protocols.

Comment:

Without question, LC-MS (MS) or perhaps LC-NMR are vastly superior methods, but they belong to a different world.

Conclusion

The possibilities of testing the peak homogeneity pointed out above can be carried out with minimum fuss, which is important nowadays where time is always at a premium.

Tip No. 32 Quick check on peak homogeneity

Part 2

Problem/Question

You are in the final stages of developing a method. Your peaks look well defined and symmetrical, the peak resolution appears to be satisfactory. How can you check without too much trouble that all components present in the sample show up?

Solution/Answer

Clearly there is a variety of possible solutions, ranging from modifying the chromatographic parameters (e.g., pH value, temperature, stationary phase) to coupling techniques such as LC-MS. Depending on your particular situation, the importance of the sample, etc., one method may be more appropriate than another.

A suggestion that can be put into practice with relative ease is as follows:

Repeat the separation using a combination of column, eluent and wavelength that is as different from the previous one as possible. The more variation in the parameters of the set-up (orthogonal conditions), the more revealing the results will be.

Here is an example:

You have just carried out a separation in an acidic acetonitrile/phosphate buffer mobile phase using a hydrophobic endcapped RP phase at 230 nm. Now is there perhaps a colleague in your lab who is using a polar RP phase (e.g., an embedded phase or some non-endcapped older material) and a methanol/water eluent at 260 nm? This is the time to put on your sweetest smile and ask him or her if he or she would mind if you used his/her set-up. Otherwise, go to the bottom of your drawer of old columns, dig out an old LiChrospher or Spherisorb ODS 1 and take a 50/50 methanol/water mixture. Flush the column at a flow rate of 2 mL min^{-1} for 10 to 15 min and then inject your sample. See also Tip No. 4. The probability that two or more components behave identically under different chromatographic conditions and at different wavelengths is relatively low. Therefore, you will get two quite different chromatograms. Figure 32-1 shows a simple example involving only two variables (column, wavelength).

Many other variables can be used in a similar way: different detector, different sample solvent, different pH value etc.

Conclusion

Two injections under orthogonally varied conditions are a convenient way of double-checking peak homogeneity and/or selectivity.

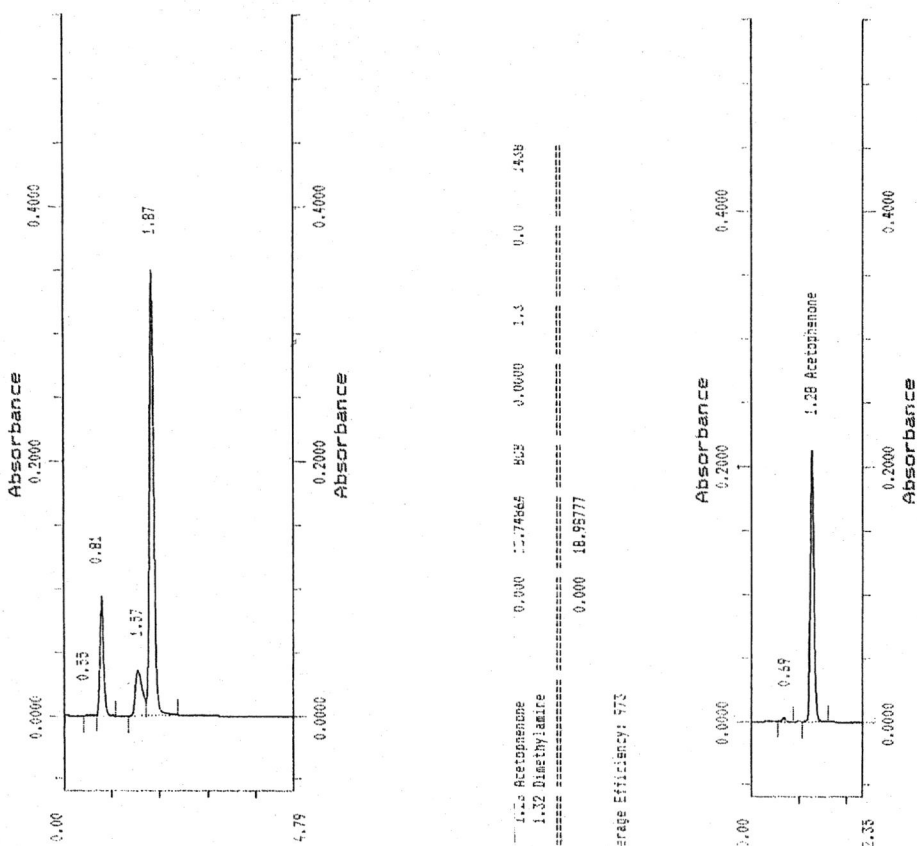

Figure 32-1. Separation of acetophenone and impurities. Left: column, Spherisorb ODS 2; λ, 200 nm, eluent 60/40 methanol/water. Right: column, Hypersil ODS; λ, 250 nm; eluent, 60/40 methanol/water.

Tip No.	**Tied to a standard operating procedure –**
33	**how can a bad separation be improved further?**

Problem/Question

Suppose you work in quality control and are tied rigorously to your operating procedures. One of them requires a resolution >1.5, but there is no way you can achieve this. Is there any perfectly legal way of doing this?

Solution/Answer

Unfortunately, not all validated, long established methods deliver optimal peaks. You can experience much in everyday life... I admit I saw the chromatogram in Figure 33-1 a couple of times shortly before Christmas (upper) and similarly only when a certain type of boss suddenly appears from behind me in the lab (lower). With great sincerity I said that I didn't know how I could improve the peak shapes – but what can you do in simple cases?

The specifications in the SOP determine your options. If everything is prescribed to the smallest detail, including the composition of the eluent as well as PC settings, your only option is to try using a new column. If there is a little more flexibility in the SOP, i.e., only the most important parameters such as column, eluent, temperature, flow, wavelength and injection volume are laid out, you would have some other options. Some of these are given below:

- Minimize the dead volume of the instrument using 0.12 mm capillaries, and perhaps you could also have a look at the connections between the various pieces of tubing or to the column. This could be quite effective. See Tip No. 25!
- Tie two or three knots into the capillary between the column and detector. This sounds like a joke, but the point is that a knot destroys the laminar flow profile, and therefore counteracts band broadening. See Tip No. 71.
- Inject between 10 and 15 µl of air with the sample. This air cushion ensures that the sample will not be diluted on its way from the injector to the column. In fact, some injectors produce such air segments on a regular base. When the air bubble reaches the column, the pressure built up in the column will cause it to be dissolved in the eluent and it will not cause any problems. See Tip No. 35/1. Consider using a restrictor capillary behind the detector. In this way, the increased pressure will ensure that air bubbles that could be in the eluent remain there and do not escape into the detector cell, where they could cause air peaks and spikes.
- The following trick can probably not be used in quality control, but keep it in mind, just in case. When injecting your sample, add some guanidine, thiourea or glycerol. This will act as a highly viscous stopper, that is pushed in front of the sample zone and prevent dilution. The good new is that guanidine elutes with the solvent front and will not affect the chromatogram.
- Dilute the sample in a solvent weaker than the eluent. If for example you are using an RP system with a C_{18}-column and a 60/40 ACN/H_2O eluent dilute the sample

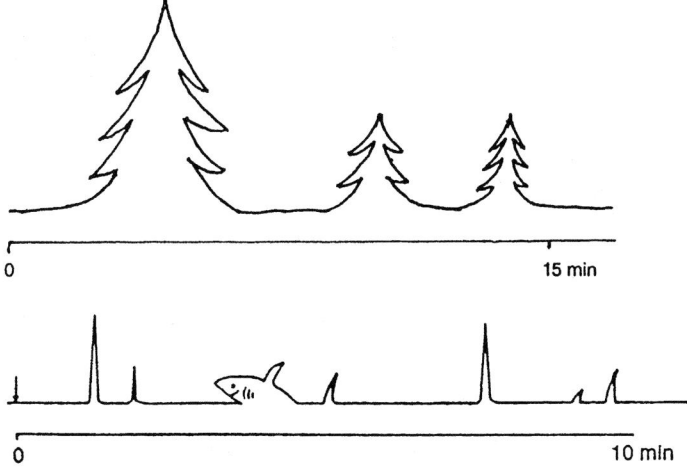

Figure 33-1. HPLC is good for a few surprises ...

in 30/70 ACN/H$_2$O. This will concentrate the analytes onto the head of the column, resulting in narrow peaks. See Tip No. 24.

All of these measures lead to narrow peaks but do not affect selectivity, the one thing you are not allowed to change. The result is better resolution through an increased number of theoretical plates, while the selectivity remains unchanged.

If you are lucky enough just to have to stick to the eluent and the column, you have the following additional options:

- Reduce the injection volume.
- Reduce the flow rate (for 3 µm particles, the improvement will be barely noticeable (and for 5 µm particles minimal).
- Change the temperature (depending on the mechanism, the change in the resolution may sometimes be imperceptible, in other cases quite substantial).

Conclusion

If you want to stick to the rules in strictly regulated austere environments such as quality control, certain monasteries or dictatorial regimes and you want to achieve sensible results within the legal limits, you have to be creative and use your imagination.

Tip No. 34 — More elaborate measures to check peak homogeneity

Problem/Question

In the previous tips, we discussed quick ways of checking peak homogeneity after a separation, leaving the chromatographic system more or less unchanged. The measures we will now discuss are more labour intensive.

Solution/Answer

1. Modifying the eluent

- Without any doubt, changing the pH value is the most effective method of changing the selectivity of ionic or even polar analytes. Suggestion: change the pH value by ±0.5 or ±1 pH units while keeping everything else constant. A second relatively simple option would be to keep the original pH value of the eluent the same, but to use a different acid or base additive, e.g., replace phosphoric acid (phosphate buffer) by perchloric acid (perchlorate). Finally, you could change the buffer ion – use ammonium instead of potassium.
- A well known classical trick is often successful – increase the proportion of water in the eluent by 5 to 10%.
- Keep the percentage of the organic phase in the eluent constant, e.g., at 50%, but replace 5 or 10% of it with a different organic solvent.
  Some examples:
  50% H_2O/50% ACN, replaced by: 50% H_2O /45% ACN
                                                                        5% THF
  or 50% H_2O/50% MeOH, replaced by: 50% H_2O/40% MeOH
                                                                       10% butanol
  etc.
  Or you simply replace acetonitrile with methanol, keeping the elution strength the same. The peaks will probably broaden, but the selectivity is quite likely to improve.

2. Changing the temperature

If you have a column heater/cooler, try to run the separation at a lower temperature, e.g., at 15 °C, or even 10 °C. Here, also, the peaks will broaden and selectivity will improve, particularly where ionic interaction and steric interactions are involved. Figures 34-1 and 34-2 show the effect of an increase in flow and a simultaneous reduction in temperature – both by a factor of 2. At a temperature of 20 °C and a flow of 0.5 mL min^{-1} 14 peaks appear, two are just about separated (look at 19 min, Figure 34-1). At 10 °C and 1 mL min^{-1} there are 16 peaks, the two problematic peaks are separated at the baseline (look at 20 min, Figure 34-2).

3. Changing the column

- Hardware
  In isocratic runs, use a longer column or smaller particles. The increased number of theoretical plates often results in a more acceptable resolution.

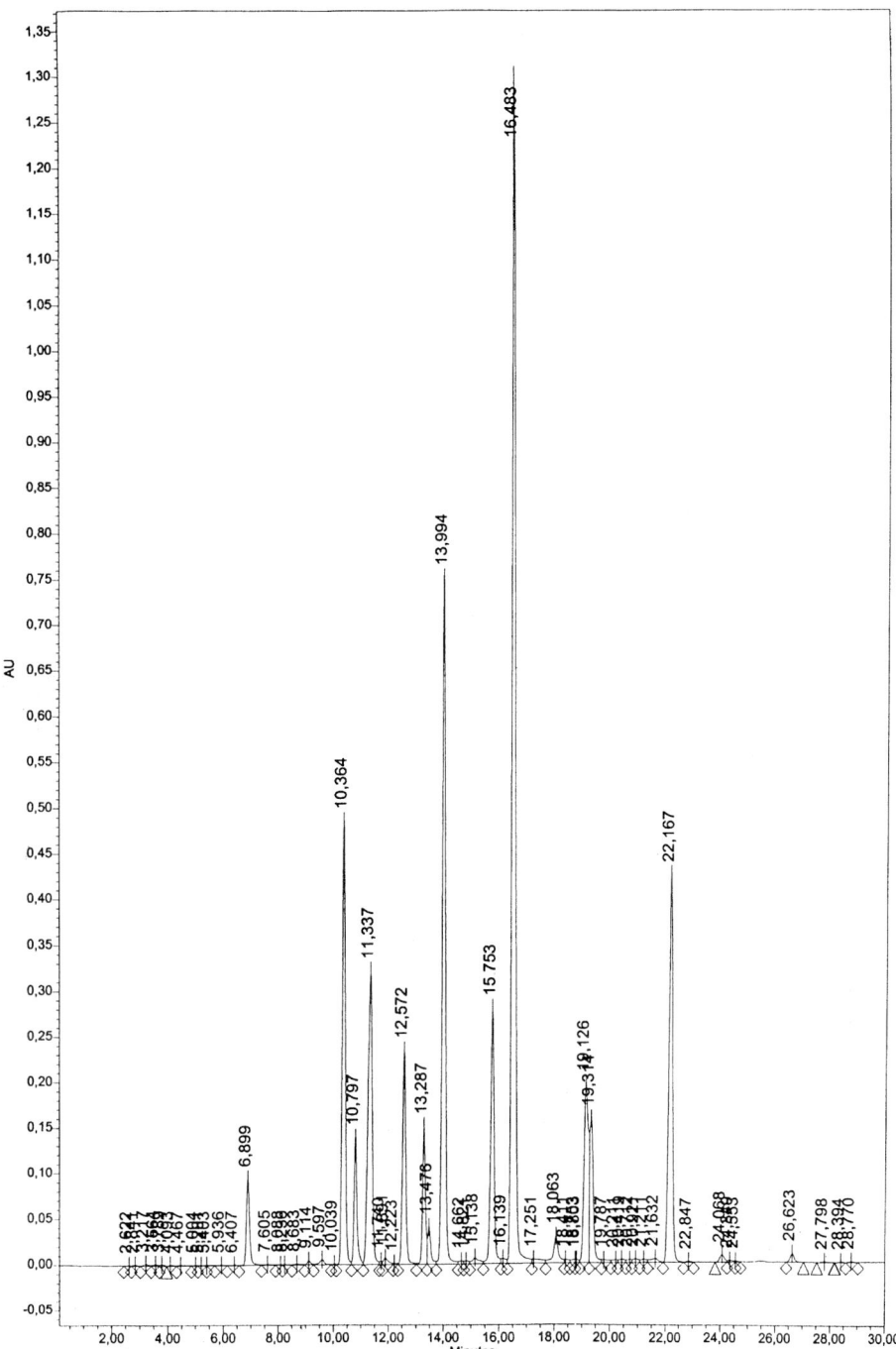

Figure 34-1. Gradient separation of a mixture of different components at 20 °C and 0.5 mL min^{-1}. Comments see text (Source Walter Nussbaum, Pfizer, Freiburg, Germany).

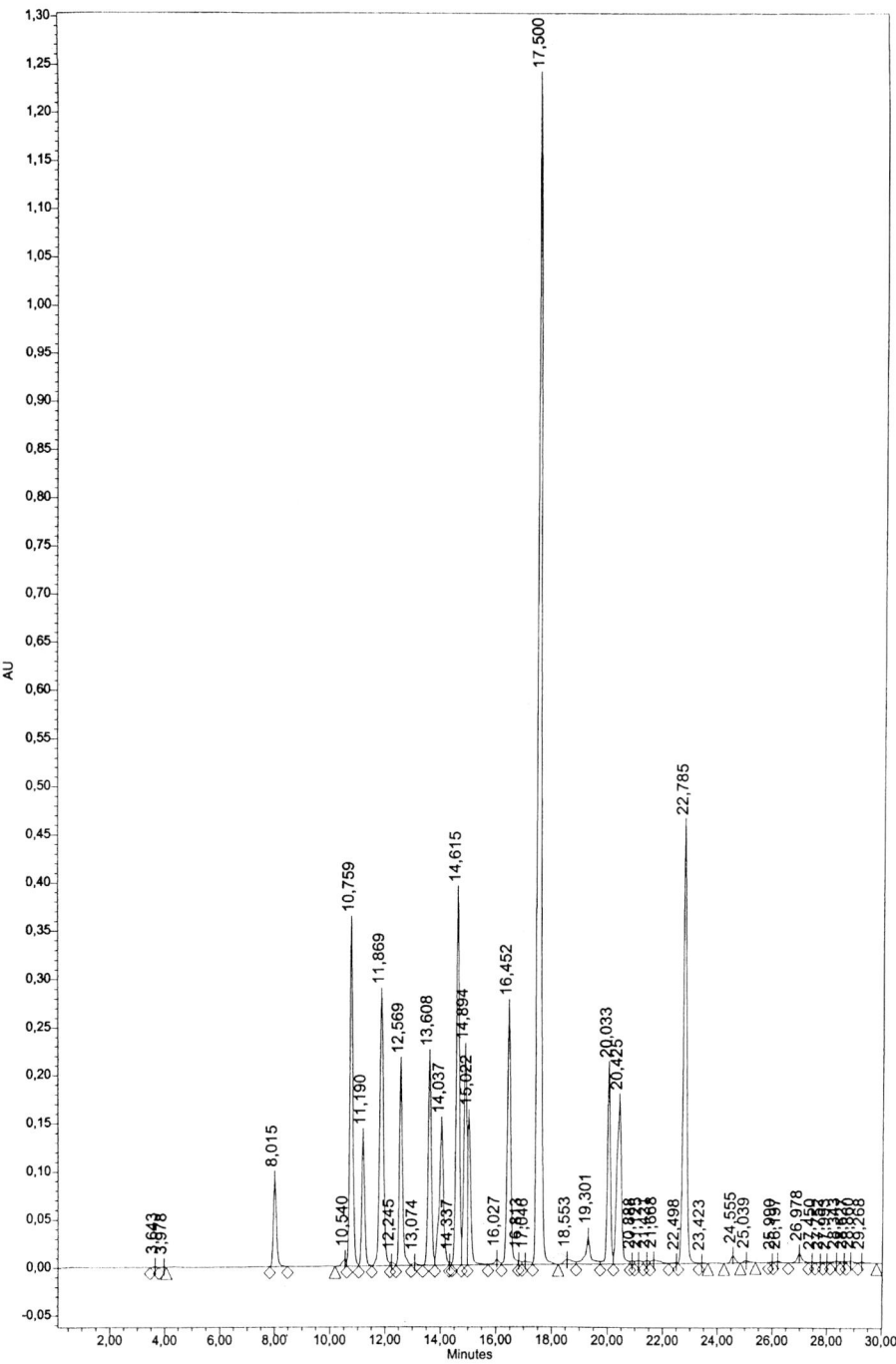

Figure 34-2. Separation as in Fig. 34-1, this time at 10 °C and 1 mL min^{-1}. Comments see text.

- Stationary phase
  As might be expected, the choice of stationary phase has quite an impact on selectivity. However, this is so complex a subject that it cannot be adequately dealt with in the limited space of this book. Read more about it in [3].

A final comment on columns – what about reviving an idea popular in the seventies, that of dual columns? Arrange two columns one after the other. Figures 34-3 to 34-5 are intended to be an encouragement. The separation on a Nucleosil C_{18} column is not brilliant, and on a CN column, it is even worse: see Figure 34-4. In Figure 34-5, both columns are now connected in series, and everything else being constant, they yield a very neat separation.

Better resolution is achieved, especially in the first part of the chromatogram. The last peak, which is non-polar, is slightly delayed by the polar CN material, but this small increase in analysis time is a price worth paying.

Conclusion

The changes suggested above can usually be achieved without going to too much trouble.

However, if you have to analyse a really important sample you should consider using tools that involve more work but yield more rigorous results, for example:

- Miniaturization down to nano-LC, see Part 2 of this book
- Column and eluent choice combined with optimization programmes

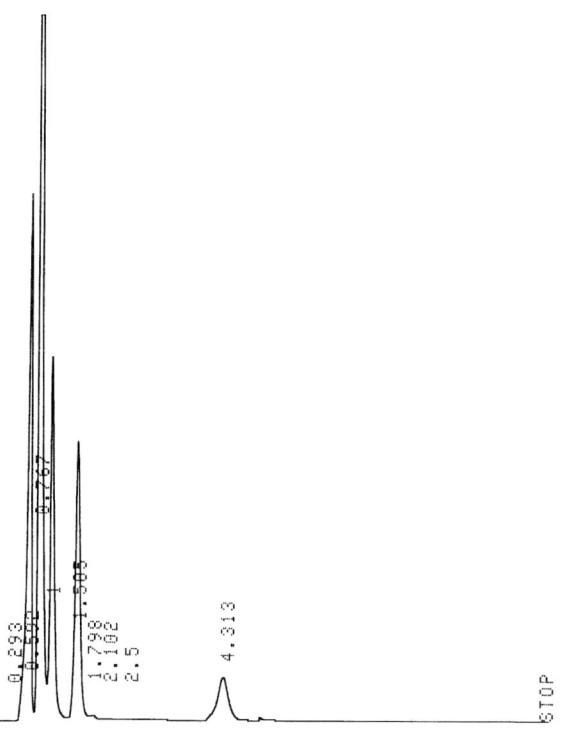

Figure 34-3. Isocratic separation of 5 peaks on a Nucleosil C_{18} column.

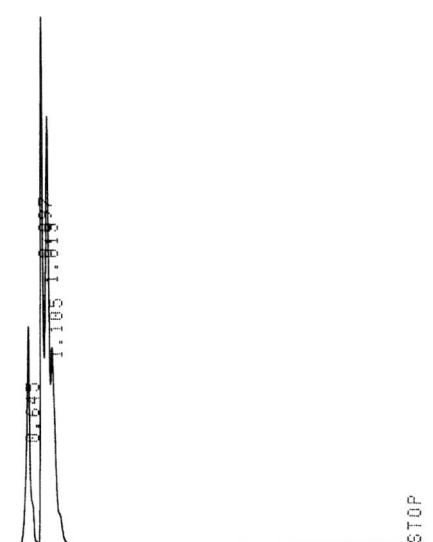

Figure 34-4. Separation as in Fig. 34-3 on a CN column.

- Coupling of chromatography and spectroscopy, e.g., LC-MS (MS), perhaps LC-NMR-coupling
- Coupling of two chromatographic techniques and subsequent coupling with spectroscopy, e.g., LC-GC-MS-, LC-CE-MS-coupling
- Coupling of chromatography and some other technique, e.g., gel electrophoresis-LC, LC-immunoassay, immunochromatography

Although the two options I mentioned last yield impressive improvements in selectivity, many users still view them with suspicion.

The third option is already fairly standard, which underlines the fact that for an important sample a sum of around € 1500–2000 € paid to external or internal service providers is by no means excessive.

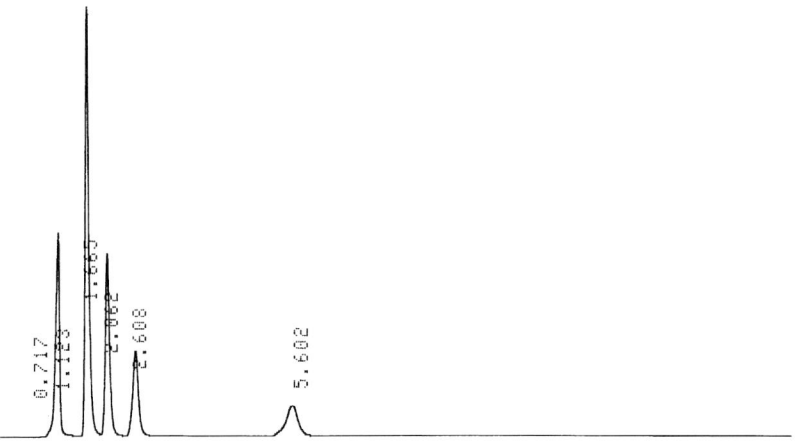

Figure 34-5. Separation as in Fig. 34-3 and 34-4 on a CN and a C_{18} column placed in series.

Dear Reader

What I have given you so far was perhaps a bit hard to digest in places. If that is the case I do apologize. May I offer you a lighter diet now? Tips Nos. 35 to 37 are short, crisp and easy to digest. Bon appétit!

Tip No. 35 — First easily digestible tip

- If you do not have a degasser and helium is too expensive, simply use (clean!) nitrogen. It is cheap and works well enough. However, you might encounter a slight problem, as some pumps tend to choke a bit more often on nitrogen. A good alternative to helium is argon.
- It is well known that an ultrasonic bath on its own is an ineffective degassing tool. However, it has been shown that in buffered eluents 5 min degassing following vacuum filtration enhances the reproducibility of retention times. Apparently, this keeps the CO_2 concentration in the buffer constant, which, in turn, keeps the pH value of the eluent also at a reproducible level.
- During routine operation, a void may form in the column, and you get double peaks. Sometimes this can be fixed by turning the column around (backflushing), Figure 35-1. Before you say goodbye to your column, you can try the following: place the column in an ultrasonic bath for approximately 5–15 min.
- If you work with labile compounds and try to dissolve them in an ultrasonic bath, you may end up with irreproducible results or ghost peaks. The natural propagation of ultrasound produces nodes and bulges. The energy reaches a maximum between the nodes. In other words, it is a matter of chance whether labile substances are totally, partially or not destroyed at all in an ultrasonic bath, depending on the location of the vial in the ultrasonic field. Whether you put a sample vial at the edge or in the centre, to the right or to the left, in metal mesh or a beaker – you will find that sometimes the water is lukewarm, sometimes hot – it is very unpredictable. If you want to achieve an even distribution of ultrasonic waves, add a tiny drop of detergent, isopropanol or higher alcohols, to the water. When validating your method in order to ensure robustness, pay special attention to the impact of the ultrasonic bath.
- Suppose you are working with an ordinary phosphate buffer. Its pH value is set at 7 or 7.5, and you notice that your column does not last very long. What could be the reason? After adding acetonitrile or methanol the pH value of the eluent shifts and is more likely to be around 8. See Tip No. 18. At this level, the silica gel begins to dissolve. If you want to extend the lifetime of your column, put a saturation column between the pump and the injector (Tip No. 07/1).
- Very clean water could have a negative effect on reproducibility and peak symmetry. Very clean water, i.e., water that is practically ion-free is aggressive. As soon as this clean water comes into contact with a glass surface, it extracts the alkali and alkaline earth ions from the glass. This makes a reproducible eluent prepara-

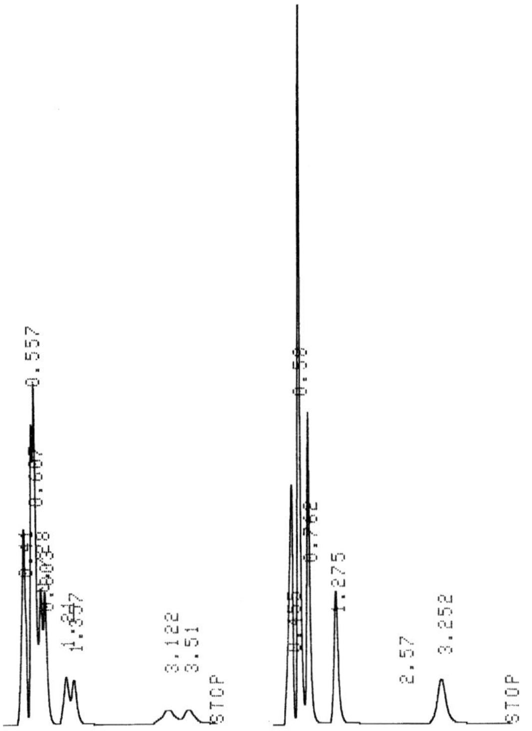

Figure 35-1. Left chromatogram: double peaks through dead volume in the column packing. Right chromatogram: removal of the problem by turning the column round.

tion difficult. Some ionic "dirt" in the water acts as a buffer and ensures reproducible results, especially in columns with older silica gel material.
- Normally you can expect to elute buffer residues from the column within 20 min at a flow rate of 2 mL min^{-1} using water or a water/methanol solution. If for some reason (LC-MS-coupling, isolation of fractions, etc.) you can't have any buffer ions on the surface of your column, it is more effective to flush the column overnight at a lower flow rate. You must, however, calculate the flow rate and elution time carefully in order to keep the water volume that flows over the column the same, as a larger volume would cause the silica gel to disintegrate slightly and thus affect the quality of the column.
- Nowadays, analytic software and HPLC apparatus often come from different manufacturers. Sometimes this causes some difficulties. If the network is overburdened, it can take 3, 4 or even 10 min for the AD converter to transfer the data to the computer. My advice is to enter 40 min even if the injector finishes after 30 min.
- Using 100% acetonitrile may prove to be tricky, as ball valves, especially those made of plastic, tend to stick because of the polymers that form in acetonitrile. There is a general risk of a polymer layer developing, e.g., at the acetonitrile inlet of a mixing chamber.

What to do:

1. Wherever possible use a mixture, even 10–15% water will do.
2. Swap the inlets A and B of the mixing chamber once or twice a month, but check whether the mixing characteristics remain unchanged, because this can be a tricky business with some mixing chambers.
3. If necessary, remove the polymer layer using THF (tetrahydrofuran).

- An isocratic separation has two main advantages over a gradient separation – it is more straightforward and thus more robust, and it yields a better resolution. In a gradient, resolution is always inferior because the distance between the peaks decreases compared with isocratic chromatography. However, the analysis time is shorter, and the peak shapes are nearly always better. Thus, a choice has to be made between robustness plus good resolution on the one hand and short analysis time plus better peak shapes on the other. You might want to opt for a compromise in the form of a shallow gradient. Simply turn your isocratic separations into shallow gradients, starting off with a gradient so shallow that it is almost isocratic, and then gradually increasing the slope until you find the optimum solution for your specific separation problem.

We are continuing our light diet of HPLC

Tip No. 36 Second easily digestible tip

- If you have to store polar RP phases, be they non-endcapped C_{18} phases, polar embedded phases or phases with short alkyl chains over a longer period of time, e.g., several months, keep them in mixtures containing aprotic acetonitrile (about 70–80% ACN) instead of methanol, which is rather polar. With methanol, hydrolysis could cause a chemical change on the surface of the phase. Some column suppliers even recommend doing this with all reversed-phase columns.
- *Ghost peaks part one*:
  Suppose some extra peaks that had not been called for turn up in your chromatogram or the peak of one component suddenly shows a hump, etc. This may, of course, have several causes, but keep in mind that silica gel has a catalytic effect. It is an excellent solid catalyst. It may well be that some of those many small peaks were not originally components of the sample but emerged *in situ* under the catalytic effect of the stationary phase. You can check this by switching off the pump while the sample is still in the stationary phase. Now your sample has time to flirt with the stationary phase – give them 10 to 20 min before you switch on the pump again. Did the number and intensity of your peaks change? This is the way to discover the secret life of your silica gel...
- *Ghost peaks part two*:
  When working in a low concentration range you discover unexplainable ghost peaks in your chromatogram. Remember that there could be interference from the environment of your HPLC instrument, e.g., a new floor is being laid in the corridor, and solvent vapours are released, or some other maintenance work is going on nearby. Is the air conditioning system frequently switched on and off? Could there be the occasional open solvent container near the ventilator of your detector? Could a security guard have used a walkie-talkie in the lab near the HPLC apparatus?
- *Ghost peaks part three*:
  You have got a very stubborn ghost peak that does not go away no matter how well you flush column and apparatus. Don't forget the degasser! If the mobile phase is contaminated, e.g., by ion-pair chromatography, some of the constituents can be adsorbed onto the membrane with its large specific surface, and then slowly desorbed again. Just to find out if this is the reason, leave out the degasser for once and run the mobile phase directly to the pump. Has the ghost peak gone now?
- *Ghost peaks part four*:
  Some acetonitrile batches don't meet the purity requirements for gradient applications. Polymers, propionitrile, methacrylonitrile, acrylamide – all sorts of things have been found in acetonitrile, a problem colleagues working on LC-MS know only too well. To solve it, add a short Al_2O_3 column to your acetonitrile solvent line. You can thus purify the acetonitrile online and at least mitigate the problem.

- *Ghost peaks part five*:
  Ghost peaks as a result of some degradation process can be really annoying. Let me just tell you about a case that happened in Switzerland. It will raise your awareness of unusual causes. As a sample was known to be extremely sensitive, extra care was taken to store it in brown glass wrapped in aluminium foil. However, it turned out that this protection was not enough. Only when the analysis was performed in a lab that had no fluorescent ceiling lights did the ghost peaks go away.
- Are you looking for an alternative to phosphoric acid in the acidic range? Think perchloric acid, acetic acid, methanesulfonic acid, trifluoroacetic acid, formic acid or hydrochloric acid. The effects on selectivity can be intriguing. Hydrochloric acid may make your pump go a little rusty if you don't give it a neutral flush-through straight after use. Well, think about it – it may be a price worth paying for what you gain in selectivity.
- Do you have to work in the pH range between 4 and 6? As you know, phosphate buffers cannot be used in this range. Here are some alternatives:
  – Acetate, between pH 3.8 and 5.8. Drawback: the UV absorption of the eluent increases, which may cause problems for a detection limit in the lower wavelength range.
  – Citrate, between pH 3.5 and 5.5. Drawback: citrate may attack steel. It is better to use PEEK capillaries. Do not leave your pump standing in this buffer over a long period of time.
  – Formate, between pH 2.8 and 4.8. As is the case for the other two organic buffers, the absorption in the low UV increases.
- Solid samples that have been kept in the refrigerator should be gently warmed before they are dissolved, i.e., they should be left to stand in the lab for about 20 min before you prepare them. The reason is that they could be hygroscopic and absorb, say, 0.1% of water. This may not matter for the first batch of samples, but by the time you have prepared the last sample it may be noticeable. You will end up with a systematic error that leads to an area reduction over time. In this context, you should also keep in mind that the volume injected is also temperature dependent. If the bottles with the standard solution and the samples differ in temperature, a systematic error due to inaccuracies in the proportions could be the result.

And here is the last of those easily digestible tips

Tip No. 37 — Third easily digestible tip

- Hexane and heptane sulfonic acid are both popular ion pair reagents, and quite rightly so. If, however, one of your bases appears too soon, you should perhaps think of camphor sulfonic acid as an alternative. Your peak then elutes later and this may facilitate your separation.
- Are you intending to separate fairly strong bases such as β-blockers or organic bases such as tricyclic antidepressants, but you only have a few older ordinary columns? Take your run-of-the-mill column and use the following eluent: 100–200 mM sodium dodecyl sulfate with about 15% 1-propanol at a pH around 7 in water or phosphate buffer, or 100–200 mM sodium dodecyl sulfate with about 10% acetonitrile at a pH around 3 in water or phosphate buffer. The chances are that selectivity and peak shape will be satisfactory.
- Does the following description sound familiar? The first peak(s) in your chromatogram show considerable tailing, while those peaks that elute later are more symmetrical. This hints at a fairly high dead volume for your instrument with this type of separation. It affects mainly the early-eluting peaks with peak volumes between 150 and 250 µL (the peak volume is the volume in which the peak is dissolved). The peaks that elute later are broad anyway, and the additional 50 or 100 µL of band spreading do not matter anyway at a peak volume of 1000 or 2000 µL. To give an example in figures – a band spreading of 50 µL ("extra column effect") is equivalent to 20–30% of the volume of early-eluting peaks. For the later-eluting peaks, the same dead volume accounts only for 5–10% of the peak volume, and the resulting broadening of the peaks hardly matters. As a consequence, the peaks elute symmetrically. Now if you are particularly interested in the early-eluting peaks, you need to lower the dead volume in your instrument. In this way, you will increase the number of theoretical plates and be able to perform more demanding separations.
- In a gradient separation in the lower wavelength range, drift can be an annoying phenomenon. There are several ways of compensating for the UV absorption of eluents A and B and thus reducing or even avoiding drift. One of them is pretty straightforward – adding a drop of HNO_3 per litre of H_2O to eluent A (aqueous phase) gets rid of the drift. See Figure 37-1 (Source: John W. Dolan BASi Northwest Laboratory).
- Check peak homogeneity (see also Tips Nos. 33 and 34)
  Your DAD seems to indicate that your chromatogram is fine, the critical peak is homogenous, the spectra on the two flanks of the peak look identical, and the match factor is over 990. Don't take all this evidence at face value! I don't want to repeat the old arguments that DAD often provides nothing more than a respectable alibi for unreliable separation results, because it is not sensitive enough at trace level and UV spectra can be irrelevant – think of the identical spectra that isomers

9 - 30% ACN/water gradient; 185 nm

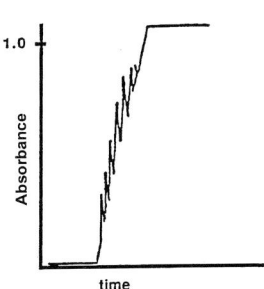

6 ppm HNO₃ in water

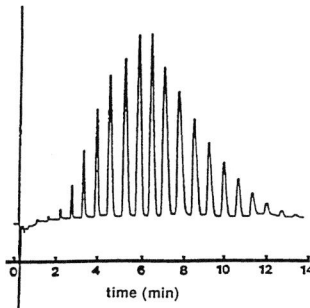

Figure 37-1. Improving drift in the lower wavelength range in gradient separations by adding a drop of HNO_3 per L to eluent A (aqueous phase) (Source: John Dolan, BASi Northwest Laboratory).

produce! Enough of that – let us concentrate on what is feasible if LC-MS or other coupling options are not available. Inject the standard substance as well as the sample at the same concentration and subtract the two chromatograms using appropriate software. Is there anything hiding on the tailing flank? Could there be a small shift in retention time? This would mean your peak was not homogenous after all, and you should look again at section 1.3 to optimize your separation.

- Are you planning to work with an eluent containing 90 or 100% water, but there is no AQ or embedded phase available? Simply use a phase with low coverage, e.g., .Spherisorb ODS 2, Platinum EPS, perhaps Platinum C_{18}, NovaPak C_{18} or Polaris C_{18}. The other thing to do is to add a pre-column or a 0.13 mm capillary in order to build up the pressure behind the column. Both measures will help to reduce or eliminate the decrease in retention time due to wettability problems.
- Do you want to separate proteins/peptides without using an ion exchanger or gel column? Very sensible! Use the following material: 3 µm, 300 Å, or alternatively, if your apparatus is suitably optimized, a non-porous material. Coverage: C_4, C_1 or a phase with perfluorinated propyl or octyl chains or simply use a monolith such as Chromolith Performance.

- Do you have a sample that is not stable in water? Keep the following mixtures in mind as possible eluents: acetonitrile/methanol, methanol/methylene chloride, tetrahydrofuran/isopropanol. *N*-butanol and dimethylformamide could be used as modifiers.
- Follow the rule that says the more polar the eluent (e.g., high proportion of water vs. water/acetonitrile mixtures) the greater the differences in selectivity due to temperature changes. By analogy, this can be applied to polar phases vs. non-polar, highly covered phases.

1.4 Troubleshooting

I am afraid your relaxation period is over – back to the grindstone!

Tip No. 38: How to approach problems in a systematic manner

Problem/Question

You work in a lab that does a lot of routine work. You are constantly faced with problems, and it would be wonderful if you always had the solution at your fingertips. However, instant solutions are hard to find in real life. If you are lucky, you may find some useful tips in the professional literature, pick up some advice from an experienced colleague or learn something useful at conferences. What I am suggesting here is a strategy on how actually to proceed. Even if it looks fairly theoretical and rather general at first glance, such a systematic approach can be very helpful.

Diagnosing and Eliminating HPLC Problems

1. Is the problem reproducible?
2. When does it occur?
   (only on a Friday, only when you are operating the instrument etc.?)
3. Write down the exact symptoms, e.g., the difference in retention time, peak area, peak height, etc., everything that changes.
4. Write down all the causes you can think of, check how plausible they are and tick the most important causes.
5. Decide which of the causes you want to investigate and what cure could be available.
6. Write a hit list of the most common problems with this method or this instrument. If possible, discuss them with colleagues, boss, method developer and supplier.

    a) Change is possible (this is rare, I am afraid).
    b) You have to put up with the situation, but at least you know it is not your fault!

Respective Comments

1. It is only worth looking at a problem if it is reproducible – a one-off does not count. You or your pump could just be having a bad hair day, or there was a huge air bubble in the system, etc.
2. Very important: Is there a correlation between a certain type of change and the symptom? This may be the first step towards identifying of the problem.

3. The most important criteria are retention time, peak area and peak height. You should take a further look at pressure, baseline drift or jumps in the baseline, spikes, etc.
4. With some simple logical reasoning, the range of possible causes can be narrowed down.
   For example:

   - If there is a shift in the retention time but not in the column dead time, the eluent composition may be the culprit, but not the flow rate.
   - If there is a change in peak area but not in peak height, the problem must lie with the pump or there may be a leak (change in flow!). But it cannot be the injector.
   - If peaks have broadened while the retention time has remained constant, eluent and temperature will be okay, but not the packing etc.
5. With the various causes in mind you can decide fairly quickly what you would like to investigate by which method and what you would like to change in the future.
6. This is the crucial and – alas – the most difficult bit! If communication between all parties involved works reasonably well, a change for the better may be achieved. If not, you will experience the stifling and frustrating effect psychological, social and bureaucratic barriers can have – not only in HPLC.

Tip No. 39 — Spikes in the chromatogram

Problem/Question

Your chromatograms look fine, the only annoying thing are the spikes that keep popping up – longer or shorter vertical lines at various places in the chromatogram. What could be the cause and what can be done to prevent them?

Solution/Answer

Air bubbles, the detector, electronics and other problems are the main causes of spikes. Let us look at these causes one by one:

1. Air bubbles
- Air bubbles in the pump (Δt_R) are easily detected – simply switch off the pump, and the spikes disappear.
- Air bubbles in the detector make a mess of the baseline. The baseline may even rise and it will take ages before it sinks again to its normal level (t_R=const.).

   Remedy: Improve your degassing and put a restrictor capillary (50 cm, 0.13 mm) after the detector

Still having problems with air bubbles? In this case it must be a leakage before the pump, or your input frit is dirty. Clean it with isopropanol, acetone or 6 M HNO_3. Finally a knick in the tubing between the solvent reservoir and pump (water jet pump effect) can lead to air bubbles.

2. Detector
- The UV lamp is too old → get a new one.
- The detector cell is dirty → clean with isopropanol and/or hot water.

3. Electronics

This would be the most unpleasant scenario, especially as the measures to be taken vary from one manufacturer to the next. It often helps if you ground only one module, e.g., the detector or PC. You could also use one single socket strip for all the modules. Using a power surge filter can be extremely effective. Also, think of shielding your interfaces, your AD converter, controller, etc., that are close to the detector. Furthermore, avoid running other large appliances such as a water-bath heater or a centrifuge on the same power circuit as your HPLC apparatus.

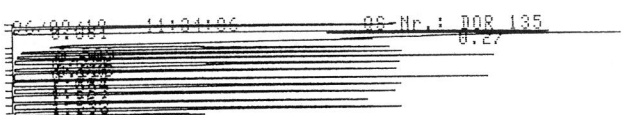

Figure 39-1. Typical diagram with air bubbles in the detector cell.

4. Other causes

Depending on the detector, the immediate or even the more distant environment can have an impact. With very sensitive detectors such as electrochemical or fluorescence detectors spikes can be caused by electric storms, trams, construction work or strong vibrations through the building. If your electrochemical detector is not sitting in a Faraday cage, it might even be that when you are all wound up after some dispute with your boss or a colleague and come too close to your detector, spikes and a disrupted baseline appear – I am serious!

Conclusion

- Once again it shows that air and dirt are the greatest enemies of HPLC. Keep them firmly out of your HPLC instrument.
- Sensitive detectors need a quiet corner and good ventilation.
- Although there are a few things you can do when the electronics go awry, let us pray that it won't happen.

Tip No. 40 — Additional peaks in trace analysis separations

Problem/Question

Suppose you work in trace analysis where absolutely every peak, even the smallest, is significant.

Let us look at two typical cases:

1. You are working with a known sample, and during the second injection, you discover two additional small peaks that are not found in the reference chromatogram. By the third and fourth injection they seem to have disappeared again.
2. You are working with a new sample. The first injection is all right, the second to fifth injections are threatening to spoil your Friday night, as new peaks show up in the chromatogram. You have spent the afternoon taking control measurements. Having looked at the chromatogram, you are now seriously thinking of moving on from HPLC to a spectroscopy lab...

What are the causes of such peaks as these that appear as suddenly as they disappear?

Solution/Answer

Naturally, trace analysis is more prone to glitches than other areas, so it makes sense to keep in mind that apart from classical causes such as power surges, air bubbles, late eluting components, etc., there may be more outlandish reasons, e.g.,

- What about solvent vapours if extraction experiments are being carried out close to the injector?
- What is the ozone concentration near the laser printer that has been printing for quite a while?
- If your HPLC is used for analysing wine, beer and other food – could there be microorganisms growing?

Let us look at the reasons for the cases mentioned above:

1. There is dirt or some solvent residue (from washing the needle) on the injection syringe. It will go away after a few injections.
2. When the septum of the sample vial is punctured for the first time, its PTFE layer is destroyed, and the sample solution is in direct contact with the septum material from the second injection onwards. This may result in small peaks – probably additives in the septum – as the residue of the needle wash solvent has the ability to dissolve components of the septum.
3. The metal syringe can have a catalytic effect, initiating or accelerating reactions in the sample solution. Thus, after a few hours or on the following day, peaks will appear that had not been there in the first place.

Conclusion

Peaks that rise or fall slowly often hint at the instability of a sample or at saturation effects (column, steel capillaries, sealing, frits). In the cases discussed here, however, the peaks appeared and disappeared rather suddenly.

Here is a list of simple measures that can help you to get to the bottom of the problem:

1. Clean the tip of the syringe with acetone, conc. NH_3 or nitric acid, flush it with eluent and then inject.
2. After the first injection, shake the vial and inject again. This will intensify the contact of the sample solvent with the septum material, and if interaction between them is the problem, you will know very soon.
3. If technically possible, inject with and without the septum.
4. Change the composition of the washing fluid in the injector.
5. Leave the metal syringe in the sample solution overnight. Inject again the following morning and compare the chromatograms. Have the areas of certain peaks increased?

Tip No. 41

What causes a ghost peak?

Problem/Question

One rarely meets ghosts in a lab – although I would not exclude the possibility that they show up sometimes to cause harm. Ghost peaks, however, are alive and well. So, what are ghost peaks? They are peaks that make an unexpected appearance in a chromatogram. Where do they come from and how can they be classified according to their origin? See also Tips Nos. 36–40.

Solution/Answer

The most frequent causes for ghost peaks are:

- Impurities in the eluent, an excess of ion-pair reagent, etc.
- Late-eluting substances
- Memory effect through desorption of a substance from the injector seal, the stationary phase, a fitting, a frit, etc.
- Air bubbles in the detector
- Degradation products of an unstable component

The first question you need to answer is whether anything in your method has changed. Don't forget the changes that may at first seem irrelevant, such as:

- A new supplier of the chemicals used to make up the eluent (salt, ion-pair reagents, other additives)?
- Did you make the buffer stronger, so that impurities in PA quality salts only appear now?
- Have you used a new pH electrode (or simply a new diaphragm, fresh solutions) and you measure the pH value of your water-based phase and then place the electrode in subsequent eluents?
- Is the water different? Has the water quality deteriorated?
- Was the matrix slightly modified?
- Different detergent used for the glassware?
- Any changes in the transport and preparation procedures or in the storage facilities and conditions?
- Change in the pH value of the sample solution?
- Did you clean the syringe of the injector lately? Perhaps you did not quite get rid of the acetone, ammonia or nitric acid you used.

If the ghost peak is a component of your sample or the eluent, it is relatively easy to find out when it entered the system.

The ratio of retention time and peak width $\frac{t_R}{w}$ is constant for a substance in a given isocratic system.

Determine these correlations for some or all conceivable ingredients of your sample mixture and of your eluent. Let us take an example – a substance that elutes after

6 min with a peak width of 2 min; $\frac{6}{2} = 3$. Suppose your ghost peak has a width of 6. If it is the substance in question, it must have begun to wander through the system 18 min ago $\left(\frac{18}{6} = 3\right)$. Did you make an injection 18 min ago? Or was it a gradient step, a change in flow or a column switch – in other words was there anything happening 18 min ago that might have caused the desorption of this substance, e.g., from a seal?

Conclusion

This method to find out where a ghost peaks comes from works quite well. If you are frequently confronted with ghost peaks try the following quick-fix procedure:

- Flush the instrument – with the column – with a solution of isopropanol, 3% H_2O_2, or alternatively with: acetic acid, water and H_2O_2.
- Clean the injector, including syringe and needle (see above).
- Passivate the whole system (without the column!) using about 30 mL 6 M HNO_3, then flush it with water to neutral.

Tip No. 42 Ghost peaks in a blank gradient

Problem/Question

It is general practice to run a blank gradient before analysing samples in gradient runs. If this test shows a significant amount of drift or even ghost peaks you need to act. How can we establish the source of the contamination?

Solution/Answer

1. Allow for a longer equilibration period before your next injection.
   If the previously found ghost peak increases in size or the drift increases, the aqueous phase is the culprit. Check the purity of the water and of possible additives. If algae in the water are a problem, use UV light. Replace the cartridges in your Milli-Q system, try out some different water, or use a different type of detergent for your glassware. Are you using a new piece of tubing in your water line? Incidentally, the flushing process seems to cause quite a lot of headaches. If the peak remains the same size, then the contamination is from the acetonitrile.
2. Does the ghost peak decrease in size during the second run?
   Then the column or some other part of the apparatus is contaminated. The answer is to flush for longer and/or more thoroughly.
   Note: If the contamination is not in the column, it is very often in the syringe or injection block of the autosampler.

Recommended procedure

- Run a blank gradient with "zero injection".
  If a ghost peak appears, clean the capillaries and detector cell.
- Blank gradient with eluent injection.
  If a ghost peak appears, clean injector parts, using, depending upon the sample, conc. NH_3, 6 M HNO_3, acetone, isopropanol, tetrahydrofuran, dimethyl sulfoxide, hexane or ether. This will drastically reduce the chances of any contamination clinging to any surfaces.

By the way – ghost peaks often appear when you have just replaced a seal, but they usually disappear again very quickly.

Conclusion

It makes sense to look for the source of what appears to be a contamination. Tracking down the culprit module and cleaning it can save having to flush the whole system with acetonitrile.

Tip No. 43 — Strange behaviour of a peak. What could be the cause?

Problem/Question

In repeat measurements, nearly all peaks behave as expected – retention time, area, peak shape, etc., remain constant. There is just one peak or maybe two that worry you, perhaps because it is constantly on the move or it does not show the area you expect it to show. What could be the reason?

Solution/Answer

I can only give you some answers, as this is one of the most troublesome problems we have to deal with in our daily HPLC routines, and I am sure that the more experienced amongst my readers could come up with any number of additional reasons.

Change in retention time:

If one or more peaks begin to wander, it could mean that there is a complex interaction between the compound in question and the stationary phase, as is nearly always the case where ionic substances are involved. Something in the chromatographic environment must have changed in such a way that only this component shows a noticeable reaction, e.g., a subtle change in pH value, saturation of the free silanol groups at the C_{18} surface, wettability problems of the C_{18} pores if the proportion of water in the eluent is high (in the meantime this is a hypothesis that is disputed) etc. Some packings with embedded polar groups contain residual amino groups that are not very stable. If one of your analytes interacts with them, you could get a drift in retention for this compound only.

Change in peak height and area:

1. Partial (physi)sorption on steel surfaces, seals and frits
   Did you change anything in your hardware or method, e.g., are you using a steel instead of a PEEK capillary, different sealing material, a different or a new pre-column, different or new septa or did you prepare your sample differently (water, pH value, solvent)?

2. Change in UV absorption
   - If the UV absorption of the component in question is pH dependent, even a marginal pH shift can result in larger or smaller peaks. See also Tip No. 19.
   - If there is just a small shift in wavelength you may be measuring on a flank of the UV spectrum of the compound in question.

3. Change in the refractory index of the eluent
   A small impurity in the eluent can cause a change in its refractory index. Now the refractive index of the eluent is not so different from the RI of eluent plus sample, so the area becomes smaller. If at all possible, it would be worth measuring not

only the pH of the fresh eluent, but also its refractory index. If there is a problem, it can be compared with the known values.

4. Sample unstable

Ask yourself basically the same questions as in 1. Did you change anything about the conditions or the hardware that could have had an impact on the stability of the compound?

Conclusion

Particular symptoms can be associated with particular causes. It is therefore important to be absolutely clear about whether certain changes affect all the peaks or only one or a few. Such problems have mainly been observed during the separation of ionic substances. Even minute changes in the method or hardware should always be considered as possible causes.

Tip No. 44

When could one expect a change in the elution order of the peaks?

Problem/Question

When carrying out an optimization process you may change, for example, the percentage of water in the eluent, its pH value or the temperature in order to achieve a change in retention times. Such optimization measures may lead to a reversal in the elution order, so peaks may be mixed up as a result. When exactly does this happen?

Solution/Answer

We can expect a reversed elution order whenever we change a parameter that has an impact on the interaction between the sample and the stationary phase.

This applies when the following parameters are changed:

- **Stationary phase**
  - Manufacturer, see Figure 44-1
  - Degree of coverage
  - Age
  - Treatment
- **Eluent**
  - Elution power
  - pH value
  - Buffer strength, see Figure 44-2
  - Type and concentration of modifier
- **Temperature,** see Figure 44-3

This is "chemistry"

A change in the sample solvent can also lead to a reversed elution order.
There is no reversed elution order if you change the following parameters:

- Flow
- Length, inner diameter of column
- Particle size

This is "physics"

Figures 44-1 to 44-3 each show one example of the three causes mentioned/changes in the stationary or mobile phase or in temperature.

Conclusion

Remembering this might save you from drawing the wrong conclusions, especially when working with unknown samples. Incidentally, if you cannot get hold of pure substances to perform addition experiments, the most straightforward criterion for assigning peaks to substances is the comparison of peak areas because the peak area remains constant, no matter when the peak elutes, as long as flow and injection amount remain constant. Exceptions to this rule are very rare indeed. One exception is a large change in the pH value.

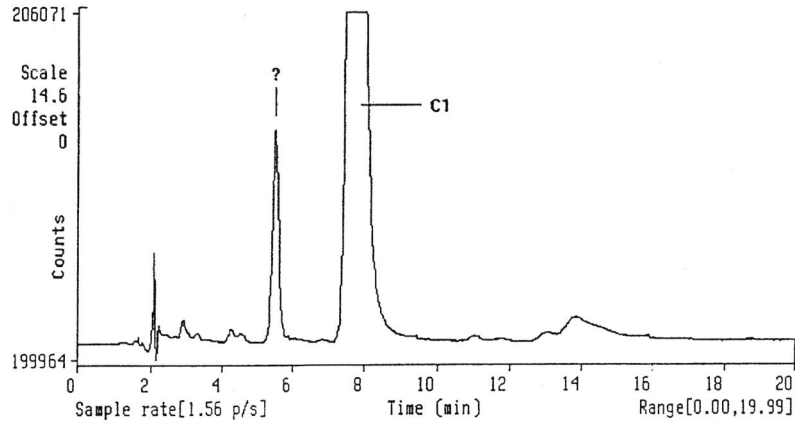
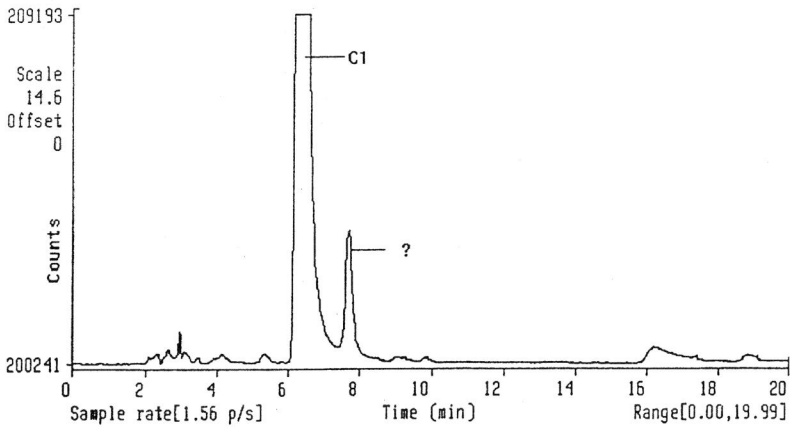

Figure 44-1. Reverse elution order on two RP-C_{18}-stationary phases.

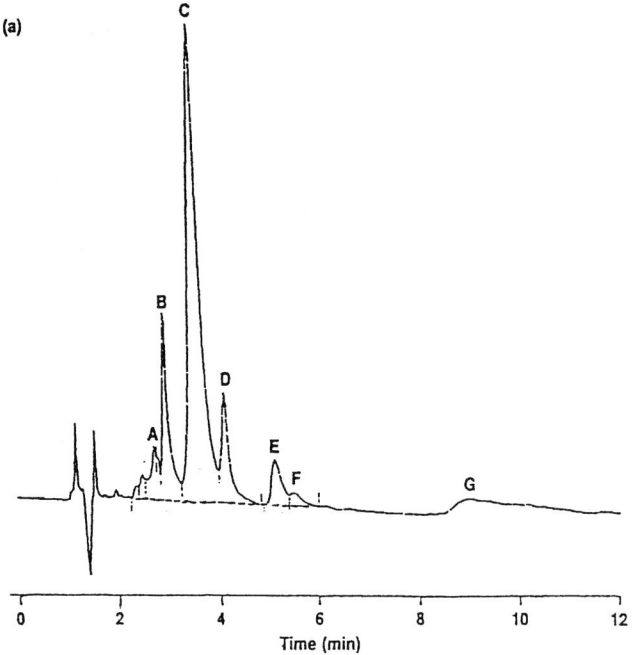
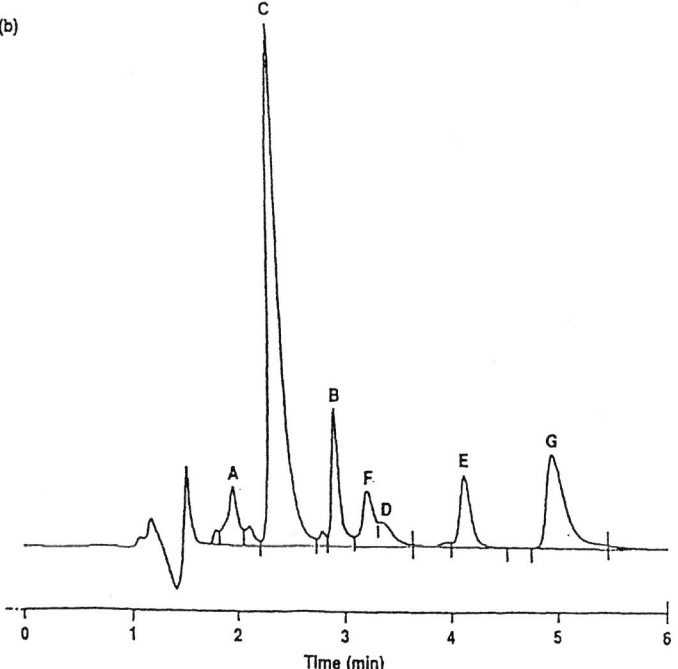

Figure 44-2. Reverse elution order due to a change in ionic strength in the eluent. (Source: Lloyd Snyder, Orinda, USA)

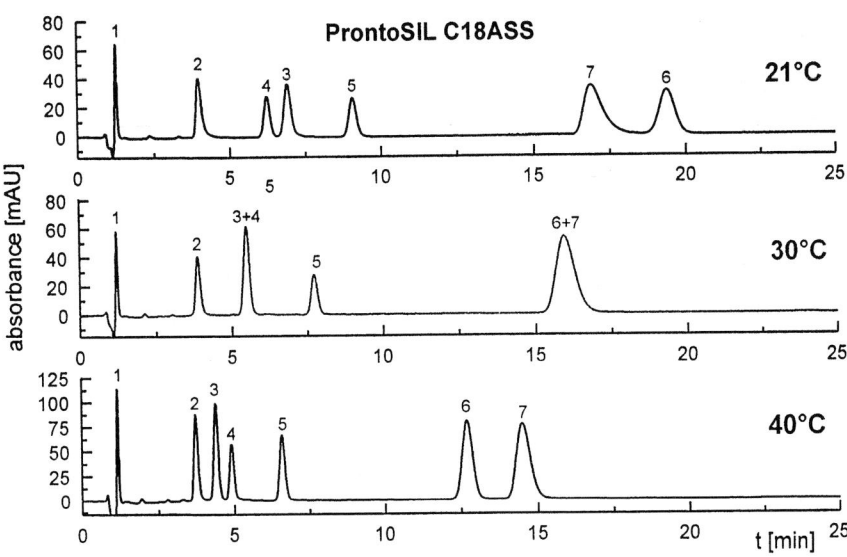

1 uracil, 2 propranolol, 3 butylparaben, 4 dipropylphthalat, 5 naphthalene, 6 acenaphthene, 7 amitriptyline; MeOH / phosphate buffer 20mM pH=7 65/35 v/v; column: 125 x 4mm

Figure 44-3. Reverse elution order due to a change of separation temperature, see separation of peaks 3 and 4 or 6 and 7 at 21 and 45 °C. (Source: Prof. Engelhardt, University of Saarbrücken, Germany)

Tip No. 45: Tailing in RP HPLC

Part 1: Fast troubleshooting

Problem/Question

Suppose you inject an unknown sample, and the peak tails. It has the shape of one of those shown in Figure 45-1. How can you quickly identify the cause?

Solution/Answer

The most common causes for tailing are:

- A second compound on the flank, i.e., insufficient resolution
- Column overload (when overloading the detector, you mostly get broad, round but more or less symmetrical peaks and the retention time remains constant, see also Tip No. 46)
- Chemical tailing, e.g., additional ionic/polar interaction between a base and the residual silanol groups of the stationary phase
- The quality of the packing is in decline, i.e., channels and dead volumes appear in the column

Figure 45-2 gives you a diagram that can help you find the cause within minutes with just two injections.

Comments on Figure 45-2:

Overloading of the column leading to insufficient resolution is more common than you might expect. When you have critical peaks, only inject 2–3 μL of sample, and you might be (positively?) surprised. The remaining three conclusions are easy to understand and should not need further comment. Tip No. 46 will discuss further general reasons for the occurrence of tailing and what to do about it.

Figure 45-1. Typical tailing peak shapes in HPLC.

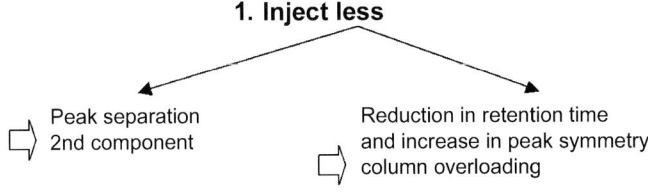

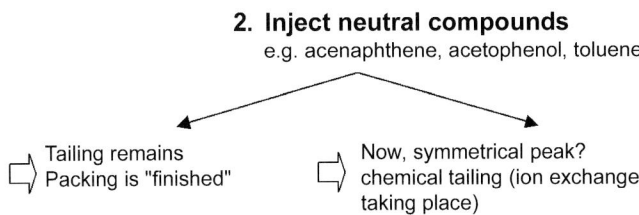

Figure 45-2. Finding the causes for tailing peaks.

Conclusion

If you have tailing don't act like a headless chicken, using a new column, changing pH value or even preparing a new eluent in a rush on the off-chance of solving the problem. Think first and use a systematic approach! This will save precious time and – even more precious – your sanity!

Tip No. 46 Tailing in RP HPLC

Part 2: Further causes and time-served cures

Problem/Question

Tailing is one of the most annoying phenomena in RP HPLC. In Tip No. 45 we looked at effective troubleshooting. Now we will investigate a few more general reasons for tailing and a few time-served methods to sort them out. See Table 46-1.

Table 46-1.

Cause	Cure
Metal ions on the stationary phase	Flush with 10 to 20 mmol EDTA
Too much organic solvent in the sample solution (rare, fronting more likely)	Dilute with water or the eluent
Deposits at the column frit	Exchange frit
Strongly acidic components on hydrophobic C_{18} phases, Figure 21-1 in Tip No. 21	Lower pH value of eluent, use polar C_{18} phase
Basic compounds on non-endcapped C_{18} phases with high silanol activity, Figure 46-1	Raise pH value of eluent, use a modifier such as triethylamine and/or ion-pair reagents, use end-capped hydrophobic C_{18} phases
Detector overload	Inject less, dilute, work at a different wavelength, Figure 46-2
Large dead volume in the instrument	Check fittings, use short, 0.12/0.17 mm capillaries
Fast early peaks in combination with inappropriate settings	Lower time constant to 0.1 s. Increase sample rate to 5 data points per second, lower peak width to 0.01 min, see Tip No. 68

Solution/Answer

Conclusion

In routine RP HPLC, the most common causes for tailing are probably declining packing quality and an inadvertent shift in pH. However, in a second step, you should also consider other causes. The environment will usually give you clues about what to look at next, e.g., separations of complexing agents (metal ions) or proteins (precipitates), an older instrument with long wide capillaries (dead volume), elution of fast narrow peaks (watch the detector settings), etc.

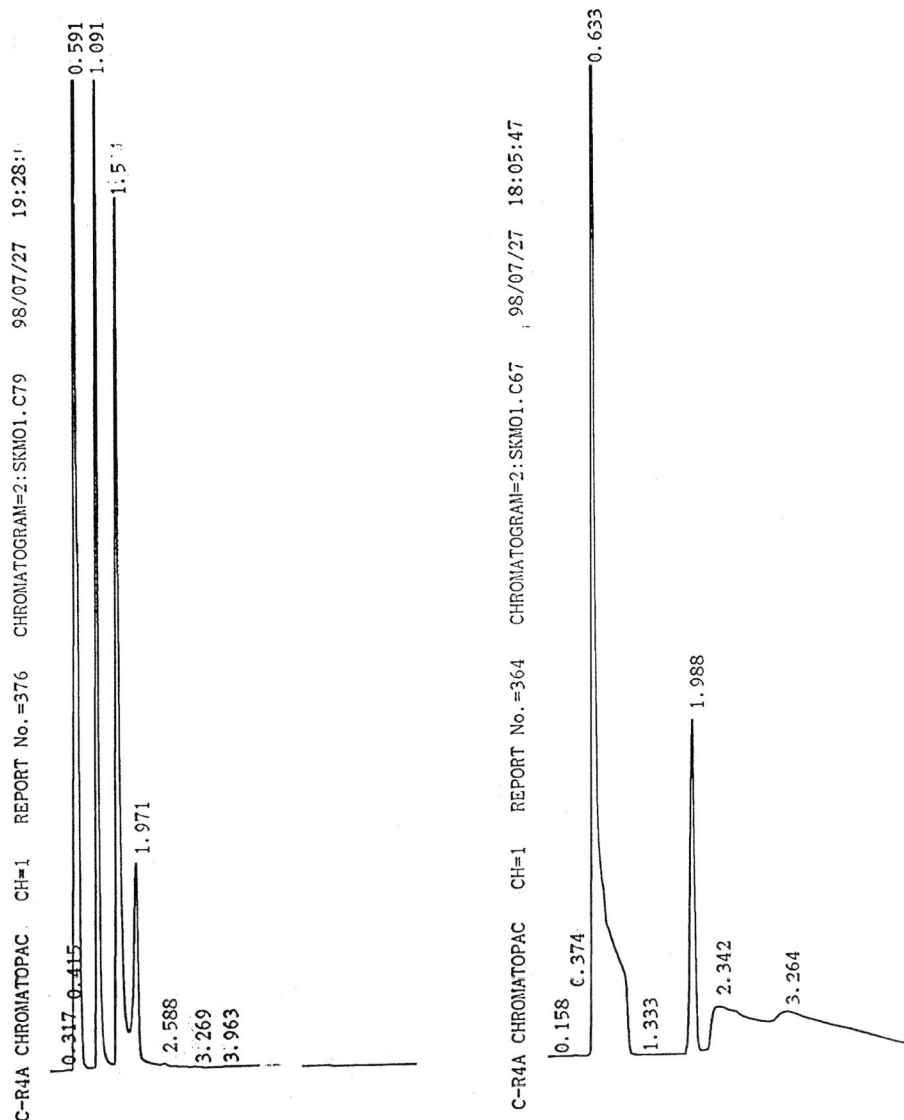

Figure 46-1. Typical peak shapes of stronger bases on a hydrophobic well-covered C_{18} phase (left) and on a non-endcapped phase (right).

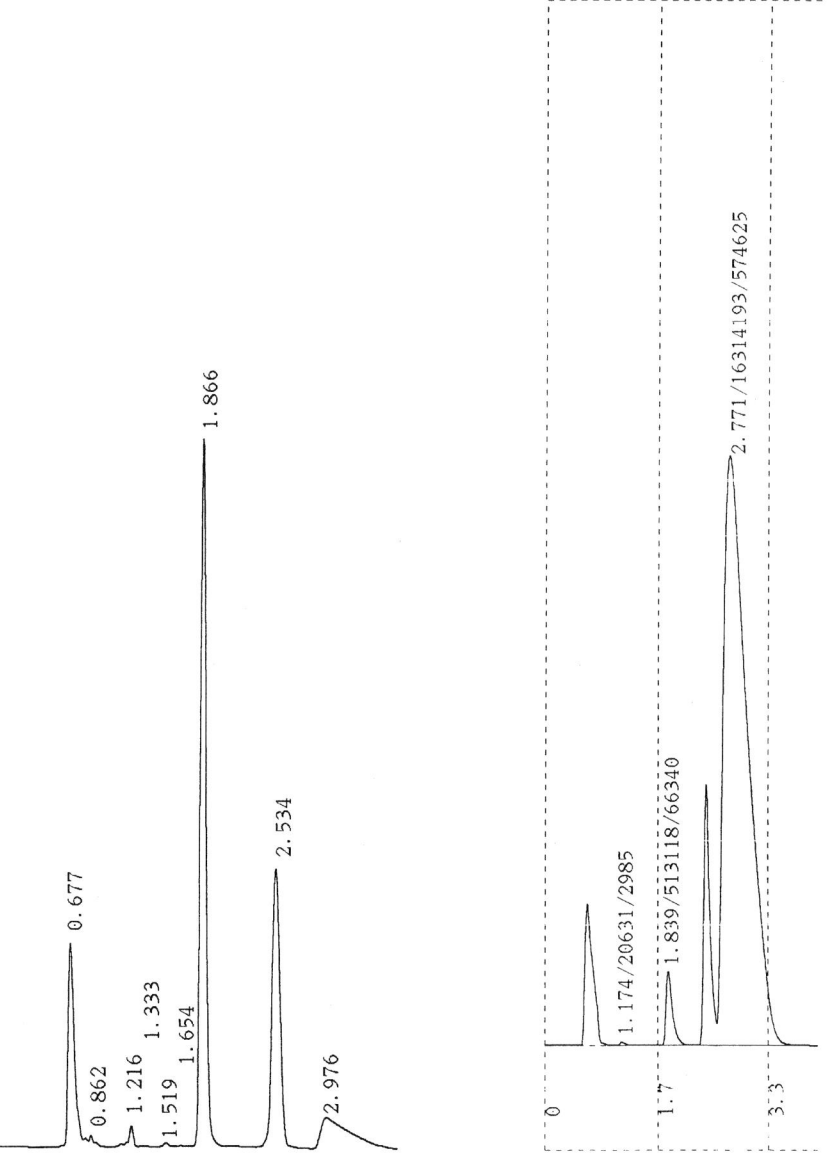

Figure 46-2. A chromatogram taken at various wavelengths. Depending on the set wavelength the peak shape is symmetrical or unacceptable.

Tip No. 47

Peak deformation and a shift in retention time due to an unsuitable sample solvent

Problem/Question

You are working with a straightforward robust method, your column is fairly new, and yet all or some of your peaks elute with fronting and/or the retention times are not consistent. What could be the reason?

Solution/Answer

Fronting (i.e., the opposite of tailing) is nearly always a certain indicator that sample solvent and eluent are not identical, or to put it more precisely, the sample solvent is stronger than the eluent. If, for example, you are working with MeOH/H_2O or ACN/H_2O or a buffer as the eluent, and the sample is dissolved in pure methanol or acetonitrile, fronting or even double peaks, perhaps also a decrease in retention time will be the result. The larger the injection volume, the more pronounced this phenomenon will be. See Figure 47-1. The chromatogram at the top shows the injection of a 1 µL sample dissolved in ACN. The eluent consists of MeOH/H_2O. The peak shape is acceptable. The chromatogram at the bottom shows the injection of 50 µL under the same conditions. Figure 47-2 proves that the ugly peak shapes are indeed connected with the sample solvent – 50 µL of sample dissolved in the eluent yield much more symmetrical peaks than the same sample volume with acetonitrile as the sample solvent. The earlier the peaks elute the stronger the deformation of the peaks – ranging from a bit of fronting or a hump (the beginning of a second peak) to double peaks. Note that this case has nothing to do with a column overload.

Figure 47-3 shows that apart from fronting and spikes near the dead time, there is a considerable decrease in retention time when the sample is dissolved in methanol and MeOH/H_2O is used as the eluent. See the left panel. By contrast, if the sample is dissolved in the eluent there are no problems (right panel).

"Different sample solvent" does not only mean that the elution strength differs from that of the eluent, but it could also mean a difference in pH value, salt content, etc.

Figure 47-4 shows that a mere difference in pH value between eluent and sample solvent can lead to major changes in the chromatogram.

Conclusion

If at all possible avoid dissolving a sample in a solvent stronger (in terms of chromatographic elution strength) than the eluent you are using. If you can't, try and keep the injection volume below 5 µL in order to reduce the shock to your well balanced chromatographic system. As an alternative, you could dilute the sample solution one- or two-fold with the eluent and then inject a larger volume of this mixture, which is more similar to the eluent. This will improve the peak shape. See Tip No. 24. A further reason for fronting could be a small pore-diameter, e.g. 50 or 60 Å.

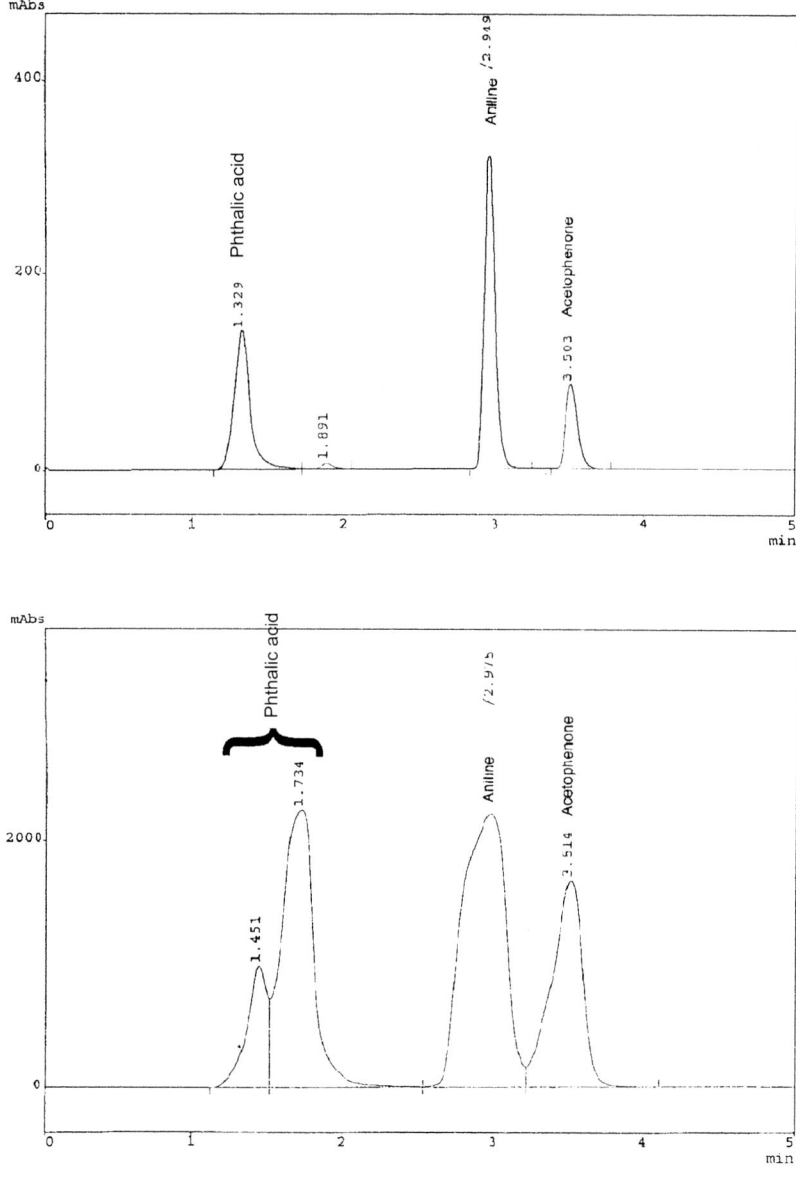

Figure 47-1. Strong sample solvent: impact of the injection volume on the peak shape. Comments see text.

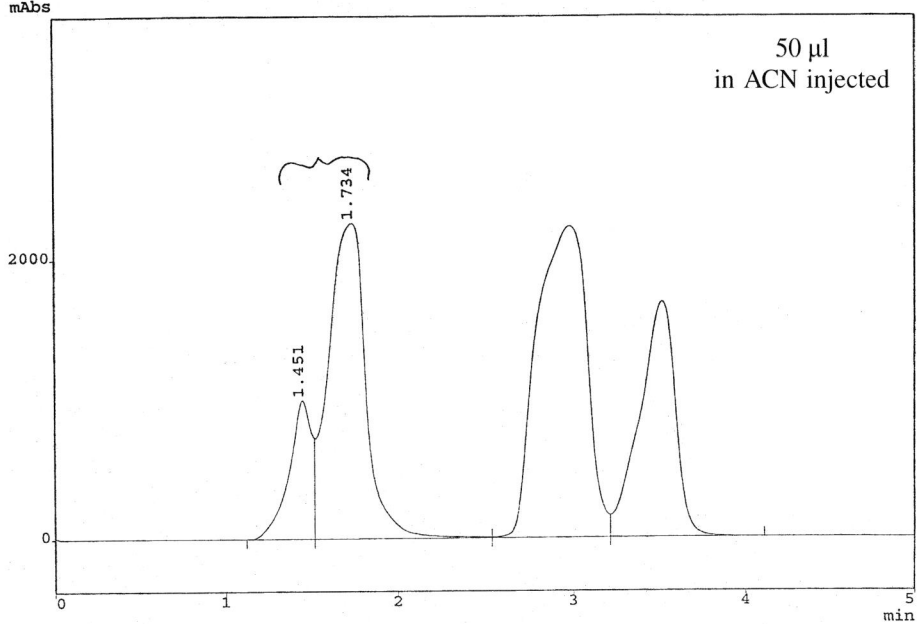
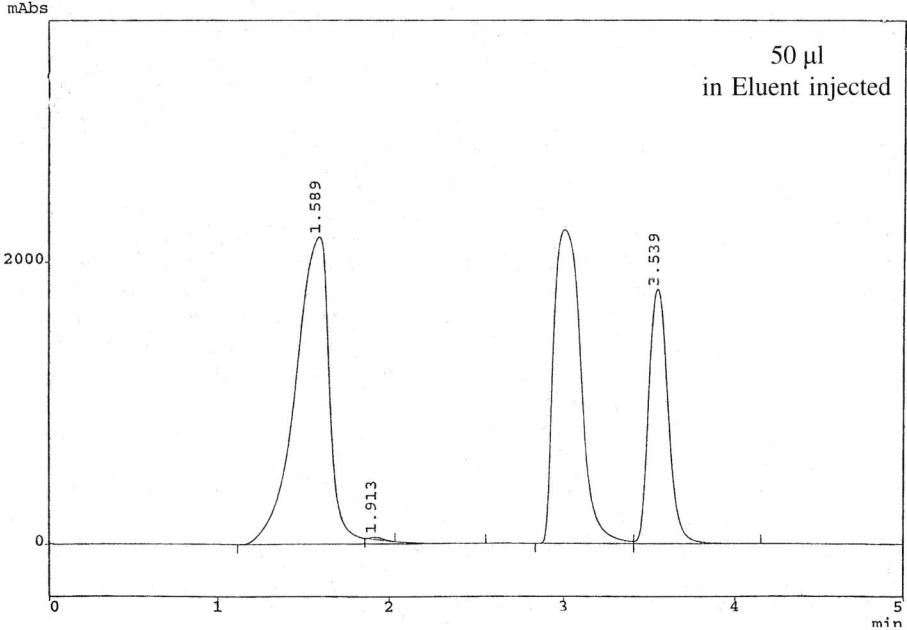

Figure 47-2. Impact of the sample solvent on the peak shape when the injection volume is larger. Comments see text.

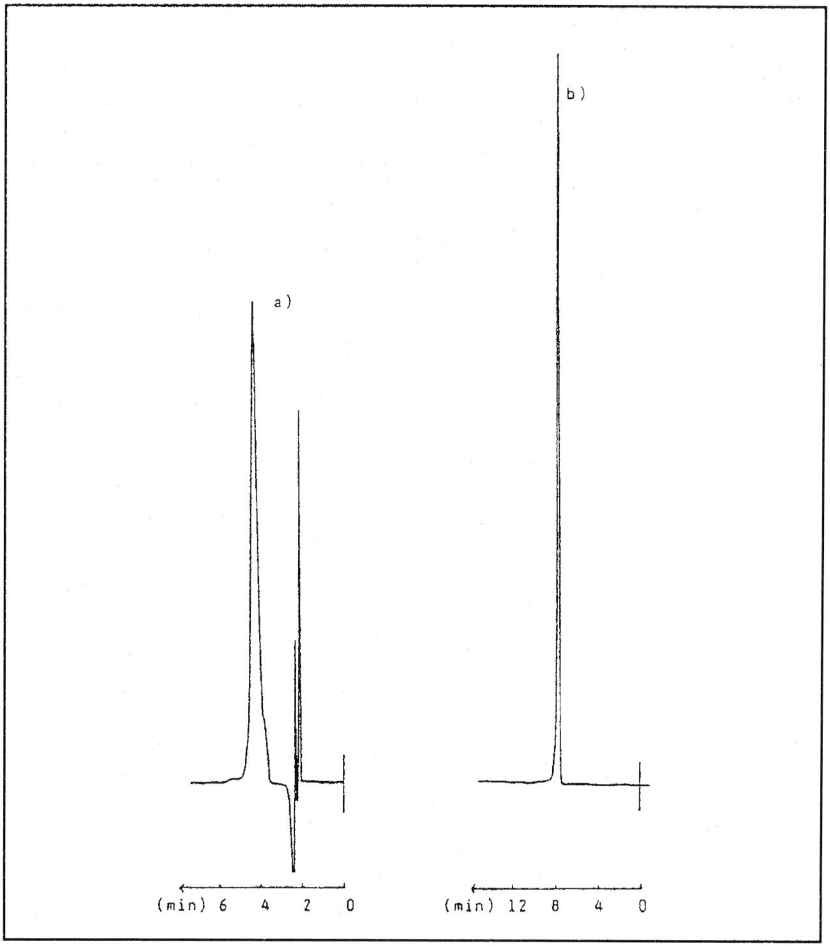

Figure 47-3. Strong sample solvent: impact on peak shape and retention time. Comments see text (Source: I. Asshauer in "Practice of High Performance Liquid Chromatography", Springer Verlag).

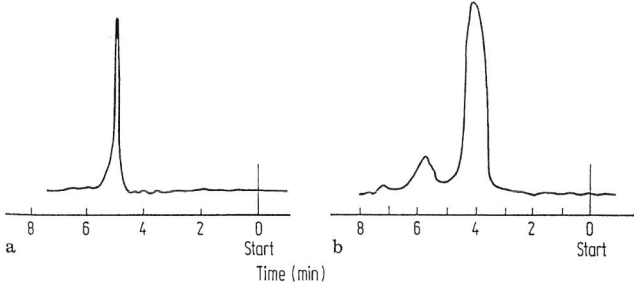

Figure 47-4. Impact of the pH value of the sample solvent on the chromatogram. Comments see text (Source: I. Asshauer in "Practice of High Performance Liquid Chromatography", Springer Verlag).

Tip No. 48

Is flushing with water or acetonitrile sufficient?

Problem/Question

Increased back-pressure, ghost peaks suddenly appearing, a broadening of peaks – all these are clear indications that there must be some contamination and you need to flush your system. Flushing seems to be an ever-topical subject in HPLC, and the necessity and effectiveness of the various steps in the process are widely discussed in HPLC courses. A frequently asked question in these classes is: "Is it enough if I flush out polar impurities with water and organic matter with acetonitrile?"

Solution/Answer

In most cases, the answer is yes. However, if the contamination is persistent it is possible that the elution power of water or acetonitrile is not strong enough, in which case one should resort to stronger eluents, see below.

HPLC labs often have their own flushing procedures in place that work well for certain samples and matrices. These are invaluable, so not only stick to them but look at them as part of your method. Even so, I am going to give you some general recommendations that have proved useful in RP columns, independent of any specific problem. There is an obvious distinction between flushing the column and flushing the apparatus.

You can also use mixtures, e.g., for inorganic contamination plus microorganisms: approximately 45% acetic acid, 45% water and 5% hydrogen peroxide.

As a rule of thumb, the amount of flushing fluid used in every step should be about 10 to 15 times the column volume. For a 125×4 mm column this would mean about 15–20 mL. By the way, to speed up the process, you should turn the column upside down.

For the instrument itself, excluding the column use (see also Tip No. 49):

- Hot water (about 80 °C)
- Hot water with a drop of detergent
- Acetone
- Diluted NH_3 solution
- 6 M HNO_3 plus a subsequent neutral flush

Special tip:
Inject the solvent you are using two or three times. This ensures that the injector is also flushed. It may sound trivial and should be common knowledge, but since so much depends on it, let me emphasize again that in the last flushing step – i.e., before you leave the column or instrument to stand for a while – the eluent should contain at least 20–30% methanol, or even better acetonitrile. This will prevent the growth of algae.

Table 48-1.

Contamination	Eluent
Normal organic contamination that accumulates in the column over time	Methanol → isopropanol → acetonitrile. Just flushing with methanol or acetonitrile should be sufficient
Stubborn organic contamination	Tetrahydrofuran (THF), dimethyl sulfoxide (DMSO), dimethyl formamide (DMF); ethyl ether is also supposed to do a good job and as a last resort, n-hexane
Salts, "normal" inorganic contamination	Methanol/water (50/50) or pure water. Pure, hot ($\sim 80\,°C$) water is very effective but may cause the silica gel to dissolve more rapidly
Stubborn inorganic contamination	0.01 M HNO_3 or 1–5% formic acid or diluted NaCl solution
Heavy metal ions	10–20 mmol/ethylenediamine tetraacetate (EDTA)
Algae, fungi, bacteria	3% H_2O_2

Conclusion

Like dissolves like – this general rule also applies here. Salts should be eluted from the stationary phase with polar solvents, organic contaminations with non-polar solvents. In many cases, flushing with isopropanol will be enough to clear the column from common not-so-critical organic and inorganic contaminations.

The measures suggested have a very good troubleshooting record. If, however, you have flushed your system in the way described and are still having ghost peaks or memory effects, and the sample, the preparation, the solvent, the degasser or the eluent are not the culprit, then your problem is not a trivial one – to put it mildly.

Tip No. 49 — Flushing and washing fluids for HPLC apparatus

Problem/Question

Parts or the whole of the instrument must be flushed when needed (e.g., in the case of ghost peaks, raised pressure), and for each purpose or module there are solvents and detergents that work particularly well. You can find a list of what works for what below. This does not include the column, which is dealt with in Tip No. 48.

Solution/Answer

Table 49-1 shows some of the effective solvents or solutions.
Final remarks:

1. No matter what you used to flush the system, before you leave it sitting idle for more than 2–3 days, you must use a fluid containing at least 20% of the organic solvent, e.g., acetonitrile in the last flushing step to prevent the growth of microorganisms.
2. More is better is a maxim that seldom applies in life, but HPLC may be an exception. It is never wrong to flush abundantly; only sometimes may it be superfluous. Usually, 30–50 mL per run will suffice.
3. Once you have flushed your instrument or done anything else to it and you want to start working again, make it a routine, even before you run a blank gradient, to inject 20 µL of eluent twice in succession. If this does not yield ghost or negative peaks or other abnormalities, but shows a steady baseline it should be all right to start work. A perfectionist would do this at 210 nm – this will show up virtually any organic contamination.

Conclusion

When it comes to flushing, the same principle as in chromatography applies. The stronger the elution power of the solvent for a given contamination, the more effective it is at eliminating it from a surface.

Table 49-1.

Fluid	Effective against/module to be flushed	Comment (where necessary)
Hot water (ca. 80 °C)	Whole instrument, even applied directly on UV cell	
0.01 M HNO_3	Inorganic contamination	
Methanol → isopropanol → acetonitrile → tetrahydrofuran (peroxide free) → dimethylformamide → n-hexane	Organic contamination	
3% H_2O_2	Algae, fungi	
20/80 Isopropanol/water	Washing of the pump seals behind the piston	In some pumps, it should also be done for buffer-free eluents (debris from the seal)
Should resemble the eluent, but not contain any salts. Its proportion of organic matter should be higher	"Purging" fluid for the injector	Using a purely organic solvent such as acetonitrile for purging can lead to fronting
Acetone, acetonitrile, methanol, 6 M HNO_3	Input frit	If the input frit is contaminated, the pump will draw air into the system
NH_3 conc., acetone, 6 M HNO_3, acetonitrile, tetrahydrofuran, dimethylsulfoxide	Syringe or whole injection block	
6 M HNO_3, acetonitrile, tetrahydrofuran, methanol, acetone	Ball valves (inline, exit line) at the pump	
Tetrahydrofuran	Inlets of the mixing chamber	Over time, polymers build up in acetonitrile. These can freeze the valves

Tip No. **50**

When does the peak area change?

Problem/Question

In quantitative HPLC, amount is almost always determined by the size of the peak area. If the changes in peak area cannot simply be associated with a difference in the injected quantity – as it should be – this will have repercussions for the results.

How can this happen?

Solution/Answer

Causes of a change in peak area
1. Change in the amount/volume injected
2. Change in flow
3. Change in wavelength
4. Change in pH value
5. Irreversible adsorption
6. Inadequate solvent

Brief comments on the various possibilities:

1, 2 and 3. These are the main instrument-related factors that can lead to an unintentional change in peak area and thus to incorrect results. This is why in a system suitability test the short time consistency of the pump, the precision of the injector and the consistency of the wavelength are tested.

4. Make sure that the pH value of the eluent remains constant, as the UV absorption of polar substances may be pH-dependent. See Tip No. 19. Furthermore, if there is a discrepancy in pH between sample solution and eluent, this can not only lead to a shift in retention time but also to a change in peak area. See Tip No. 47.

5. Larger molecules such as proteins in biological matrices can be irreversibly sorbed at the surface of steel capillaries, seals, etc. (hungry surfaces) and are not always detected. The danger is greatest if pure water is used as the injection solvent. Remedy – add a small amount of isopropanol, MeOH or ACN to the sample solvent.

6. If the sample solvent is stronger than the eluent, it will result not only in a decreased retention time and troublesome fronting, but also in a minor or major change in peak area size.

In a worst-case scenario, all these adverse effects reinforce each other – as the Hungarians say, "When the devil appears, he usually brings along his mother-in-law":

RP separation of a mixture of substances from a complex matrix, dissolved in water, by means of a steep gradient with a non-robust buffer coupled with detection on the flank using a low wavelength.

Conclusion

1. Aim for robustness when developing a method.
2. Check your instrument frequently. This does not mean you have to run system suitability tests all the time or do a daily calibration. It is far better to use control charts intelligently, which renders daily calibration superfluous [4].

Tip No. 51 — Reasons for a change in either peak height or peak area, but not in both

Problem/Question

There is a whole host of reasons for a simultaneous change in peak height and peak area, as the more experienced readers will know – such as a change in injection volume, instability of the sample, a leakage in the injector, perhaps a change in pH (see Tip No. 19), irreversible sorption on "hungry surfaces" (Tip No. 50) or a change in wavelength. Now what about a change in peak height with the peak area remaining constant or *vice versa*? Let us suppose that you have not deliberately changed around any of hardware, such as replacing a column by a longer one or by one with smaller particles.

Solution/Answer

1. Change in peak height with the peak area remaining constant

In principle, we can distinguish two reasons:

A. A change in interaction. Here the retention time also changes.
- The eluent or its pH value
- The packing material, e.g., a chemical change on the surface of the stationary phase.
- The temperature, e.g., if the column oven has been open for a while.

B. A change in the theoretical plate number. The retention time remains constant. The reason for this is:
- The quality of the packing of the column is in decline. There may be channels in the packing.

The most common causes are usually either a decline in packing quality (constant retention time) or an inadvertent change in the eluent (change in retention time).

2. Change of peak area with the peak height remaining constant

Only two reasons for this should be considered – a change in flow or an air peak under the main peak.

The cases above describe an ideal situation. In real life, there may be aberrations from the pure doctrine. The following three examples demonstrate this:

- A change in flow also entails a (minor) change in peak height.
- A change in temperature can lead to a change in pH value, which, in turn, can have an effect on the peak area because sometimes UV absorption depends on the pH value.
- The addition of modifiers to the eluent can have an impact on its UV absorption, and if the signal-to-noise ratio changes, the peak area may also change.

Conclusion

Even taking into account that there are special cases such as those mentioned above, it is well worth having a close look at the chromatogram to track down potential errors with respect to peak height or area.

Tip No. 52

Excesses and their pitfalls

Problem/Question

You are working with an ordinary linear gradient, trying to make assay analysis. Your system is reliable and you know your injector as well as your pump is in top condition. The content of some of the peaks is fine, but in some, or maybe only in one, you find a slight but reproducible and significant excess. By performing several tests you exclude all the reasons for this that you can think of. Are there any other strange reasons for this phenomenon?

Solution/Answer

I am sure there are any number of them – after all HPLC is full of mysteries! Let me give you just one:

Run a blank gradient. It is important that you can see the profile of the gradient. If necessary, add an eluent B, 0.1–0.2% acetone or some other UV modifier. Look at the profile of the gradient – is it linear or does it sag slightly? If the latter is the case, check if the problem peak happens to elute when a specific eluent composition is reached and a discrepancy between the expected linear shape and the actual, slightly concave shape can be observed. You could use a ruler and draw in the expected linear gradient, and if you find a discrepancy you may have a certain degree of decrease in the volume, i.e., perhaps the flow of the pump is not the preset 1 mL min^{-1} but only 0.97 mL min^{-1}, and this is being reproduced. This would be a typical case of a systematic error, resulting in a slight increase in the area and therefore the analysis, as the product of flow time areas remains constant. In this context the following should be noted: in case of critical measurements you should consider the compressibility of the fluids, generally 1% per 100 bar. Working at 200–300 bar and mixing solvents with a large difference in compressibility (e.g., methanol and water), you should use the correction possibilities that modern instruments allow.

Conclusion

Occasional air peaks that elute under the peak and have an effect on the area but not on the height, co-eluting impurities that are indistinguishable in their UV behaviour from the relevant component, and the case described above, these are, alas, the more difficult-to-find reasons for a lack of reproducibility if the composition of a sample is determined via the peak area.

Tip No. 53 Algae, fungi and bacteria in HPLC

Problem/Question

When water and in particular buffer solutions are left to stand there is a risk that microorganisms might grow, which could result in increased pressure, ghost peaks and even tailing. How can you prevent this growth, and if it has happened already, how do you get rid of the miniature botanical garden growing in your system?

Solution/Answer

These are the conditions under which many little beasties and plants will thrive:

- Stagnant water, sunny spots, cozy corners with hiding places, rough surfaces – in other words, eluent standing around, daylight (e.g., buffer solutions on a windowsill), frits, fittings, mixing chambers, rough, dirty surfaces in the instrument provide ideal breeding grounds.
- There is plenty of food and drink in the shape of buffer solutions garnished with carbon compounds. Chef's special – ion exchanger, dessert: sugars and dextrans.
- Summer temperatures are preferred – no sunburn please, and not too acidic or too basic, which means: 30–40 °C promotes growth, UV radiation kills most germs, and a strongly acidic or alkaline medium will prevent microorganic growth in most cases – not always, however.

This should give you some idea of what your best weapons are in the fight against microorganisms.

- Never leave a buffer to sit in the instrument over a long period of time – keep it circulating.
- Use brown storage containers wherever possible.
- If you need to store material for a longer period of time keep it in the fridge.
- Do not store buffers as pure solutions but add an organic component (e.g., more than 20% ACN) immediately.

A few extra tips are given in Table 53-1:

Conclusion

We lab analysts are in favour of protecting flora and fauna worldwide – with two exceptions – our HPLC system and our eluent storage containers.

Table 53-1.

• When storing buffer solution, use additives	Add 0.01% sodium azide to kill microorganisms, and the good news is that sodium azide elutes with the solvent front and will not spoil your chromatogram
• Keep your surfaces smooth and clean to give microbes no breeding ground	Passivate system with 6 M HNO_3
• Eliminate microorganisms from the system	Flush with about 3% H_2O_2
• Targeted cleaning procedure where needed	Clean, e.g., syringe in the injector, loop in the Rheodyne valve or input frit with 6 M HNO_3 or NH_3 conc.

Tip No. 54 Does 40 °C always mean 40 °C?

Problem/Question

We have frequently discussed the various reasons for a shift in retention time, and a difference in the actual separation temperature was one of them. Suppose you are following a protocol that stipulates 40 °C, so you set your column oven accordingly. If you now notice that the retention time has shifted a bit, you would normally not think of a discrepancy in temperature as the reason, as the temperature display says 40 °C. Beware – could it be that 40 °C does not always equal 40 °C?

Solution/Answer

I am afraid so! The actual separation or detection temperature depends on many parameters. I could easily fill several pages on this subject – but here, we want to keep it short and snappy and just concentrate on essentials.

Note: The extent of the heat dissipation depends on the immediate surroundings of the column, which means that not only the set temperature matters but also the tempering. You need to investigate the following questions:

- Is the column housed in an aluminium block or does it have a fan thermostat?
- Is the eluent temperature also set by a thermostat?
- Are you using a steel or a PEEK capillary (difference in conductivity!)?
- Are you using the same capillary diameter?
- Could there be a heat exchanger capillary in the detector?
- Is there a thermostat in the detector?
- There is a pressure gradient in the column that follows the direction of flow, which entails a reverse temperature gradient. The temperature difference can be 2–4 °C per 100 bar and depends on the polarity of the eluent. Are you really working at the pressure given in your protocol?

In an RP mechanism, an absolute temperature difference of 2–4 °C is not so critical, but if you are separating polar substances and using buffers (pH dependence on temperature) differences in temperature could show up in retention time as well as in peak shape.

Conclusion

Keep in mind that user-friendly operation protocols not only give you the parameters for the separation, but also indicate the hardware and settings you should use – i.e., not only evaluation software XYZ, but also the sample rate, not only 40 °C, but also the type of column oven, not only 0.12 mm capillary, but steel or PEEK. It should further let you know whether the eluent is also subject to thermostatic control and whether the detector used has a heat exchange capillary, etc.

Tip No. 55

The most common reason for a lack of reproducibility is a lack of method robustness

Problem/Question

In routine HPLC, insufficient reproducibility of peak area or peak height is a widespread problem. If a fault in the system has been ruled out by calibration and qualification, the cause can nearly always be found in a lack of method robustness. In other words, when methods are validated robustness has not been scrutinised sufficiently. The consequences of non-robust analysis can be dire indeed, leading to increased costs because of complaints and having to carry out repeat measurements or even discarding whole batches of substances that had nothing wrong with them. There are any number of examples to prove my point. What aspects should a person developing or validating a method keep in mind in order to ensure its robustness?

Solution/Answer

There is an extensive discussion of robustness in HPLC methods in reference [4]. You will find the major points summarised in Table 55-1.

Firstly some suggestions and comments on the procedure when checking robustness:

All critical parameters of a method should be deliberately varied within the relevant range and the impact on the result documented. This is where the analyst's skill and experience comes in. The expert must have the know-how and the time to determine which parameters are critical for the method in question. Thus, no time is wasted on unnecessary tests, and nothing important is left out.

Here are three typical examples:

- Are strong basic compounds to be separated? Vary the pH value of the eluent!
- Are proteins to be separated? Check surfaces for irreversible sorption!
- Is the method to be used as a routine tool in several labs? Check 2–3 columns of various batches for reproducible properties!

In order to narrow down the options of the causes of non-reproducibility, let us assume that there are no technical flaws in the instrument – pump and injector, for example, just passed a technical check-up.

Remember that robustness must be tested with genuine samples (analyte with placebo, matrix, auxiliary substances or degradation products resulting from stress tests). By no means should you use clean standards because the method must prove itself in real life with real samples. This is often neglected and a common reason for hiccups with a method in a routine lab situation.

Table 55-1. Robustness tests in HPLC (example)

	What can be checked?	Where can a change occur?	Comment (if needed)
1	10 or six repeat injections (perhaps replace steel capillaries with PEEK capillaries)	Area, height	Irreversible sorption of the sample or some analytes? When are all active centres covered?
2	Injection of twice the usual amount or volume	Area, height	Has the peak area doubled while the retention time remained constant? Is there a risk of column overload under these conditions?
3	Vary wavelength by ±2 nm	Area, height	Are you perhaps measuring on a UV flank and if so, what is the effect on the signal?
4	Vary temperature by ±5 °C	Retention time, height (perhaps also area)[a]	In this and the following experiment keep in mind the possibility of an inversion of the elution order
5	Elution strength; vary organic proportion by ±5%	Retention time, height	
6	pH value, e.g., ±0.5 pH units	Retention time, peak shape (perhaps also area and height)	
7	pH value of the freshly prepared eluent (after addition of methanol/acetonitrile) at the beginning and towards the end of the measurement as well as pH value of the eluate	Retention time, peak shape (perhaps also area and height)	
8	Vary ionic strength by 1±5 mmol	Retention time, peak shape, height	
9	Ageing of phase	Everywhere	After several days, another chromatogram is run, or the influence of phase ageing is continually monitored with the help of control charts. Record chromatographic parameters such as resolution, area, retention time and tailing factor in relation to the time elapsed
10	Breaking-in time of a column, calibration time in gradient separation, duration and mechanism of degassing, etc.		

a) Temperature has an impact on the pH value, which, in turn, can influence the degree of dissociation and thus the UV absorption of ionic analytes. The signal changes accordingly.

Comment

"Neither too much nor too little" is a good maxim to follow. It is recommended that you always carry out points 1–5. The work involved is within reason. If you are dealing with ionic substances, points 6–8 are a must. Checking point 9 with control charts is by far the more economical solution. Point 10 gives examples of important factors that can influence the result. For economic reasons, these need not always be varied but must be documented scrupulously. These factors must be included in any SOP, as well as the specification of the dwell-volume in gradient systems, settings such as threshold and sample rate, procedures of eluent preparation, duration of ultrasonic exposure, details of the solvents used when dissolving the sample, specifications regarding the installation of a precolumn and the quality of eluent additives (p.a. for chromatography, for spectroscopy), etc.

Conclusion

When considering projects concerned with reducing costs and increasing efficiency, I often find that inadequate or non-robust analytical methods are to blame for a cost explosion that can worry a company and its board over many years. I cannot discuss this extensively here, but it has to be said that money spent on early verification is money well spent. Careful decisions about an objective-related analysis design in quality control (e.g., a decision between DAD and the more robust, more sensitive and cheaper but not so fashionable multiple channel detector) are crucial investment decisions.

Have a break ...

Dear Reader,

You have bravely soldiered on to reach this point in the book – quite an achievement. You truly deserve a break. Of course I don't mean that you should shut the book and go away. It's playtime.

When you look at page 139, you will find the beginnings of a list of sentences. Please complete the sentences so that they make sense! If you can think of several alternatives – so much the better.

On page 140 you will find bits of sentences, beginnings and ends, all over the place. Please match up the beginnings and the ends of the sentences! You may find that the first half could be completed by more than one bit. However, you should match the halves in a way that: (a) none of them are left over and (b) **all** of them make sense.

On page 141 there will be a happy reunion with Peaky and Chromy. What do you think of what Peaky has to say? Has he learnt his lesson properly?

The answers are to be found on page 265.

Complete the sentences

Please continue the following sentences to turn them into valid statements. Sometimes there may be more than one answer.

In an RP system, ionic substances may elute very early. If that is the case I should ...

If only three out of eight peaks in my chromatogram are tailing ...

I can tell the packing quality of my column is okay because ...

When I try to separate ... on Spherisorb ODS 1 or Hypersil ODS without using additives in the eluent I will get tailing peaks.

In an eluent mixture if methanol is replaced by acetonitrile at the same elution strength, the following will definitely change:
....
and perhaps also the following:
....

Some operating protocols demand that selectivity between peak Nos. 4 and 5 be $a \geq 1.5$. This does not make sense because: ...

Although the retention time has shifted I know that temperature and eluent are okay because ...

If I want to separate basic substances in an acidic medium I should expect ...

The pressure has increased. On the basis of certain facts/pieces of information (given on the left) I can exclude the following causes (on the right).

	To be excluded as causes for the increased pressure:
The dead time is constant	...
My column oven is working correctly	...
The composition of the eluent has not changed	...
Eluent used methanol/water 50/50	...
I am working in recycling mode	...

"Matching pairs"

The area depends ...
... depends on temperature, eluent and the type of column packing.

When there is a shift in dead time ...
... selectivity (separation factor a) usually decreases while efficiency (theoretical plate number N) increases.

The resolution depends ...
... than an alkaline eluent.

When you change your C_{18}-supplier ...
... on the flow rate and the injection volume.

When you change the composition of the eluent ...
... than salt in the pump.

Cold columns ...
... there is no need to prepare a fresh eluent.

Better a 5% error margin in the integration ...
... not only the retention time, but also the elution order may be changed.

When you raise the temperature ...
... on flow, stationary phase, temperature, particle size and the dead volume of the apparatus.

If you want sharp peaks ...
... will change the retention time and peak shape of ionic substances dramatically.

Better sour cherries ...
... everything except the area can change (unless something is caught up).

Retention factor k ...
... yield hot (good) separations.

Changing the pH value.
... does not depend on the flow or the length of the column.

Selectivity ...
... than 10% in the sample preparation.

Pepper in the soup is better ...
... you must have acetonitrile in the eluent.

Has Peaky remembered his lessons correctly?

Our two old friends Peaky Acid and Chromy Silicasky met up again after a long time, this time on a giant roundabout at a fair, also known as an autosampler. They were only No. 96 in the queue, also called a sequence. Given all the usual delays at the fair, otherwise known as retention times, they had plenty of time for a chat. Peaky proudly reported that he had been listening in when chief lab operator Nicolas Pump introduced his younger colleague, Ms Cell, into the secrets of HPLC. "You know it's really exciting stuff, real science, all to do with acids and bases. Let me tell you a few things so you don't have to die in ignorance!" This is what he told Chromy who was listening with awe:

1. It makes sense to use endcapped phases to separate acids and bases. This will result in symmetrical peaks.
2. Many of the latest C_{18} phases are not so suitable for the separation of strong acids.
3. Methanol/buffer eluents are better for the lifetime of a column than acetonitrile/buffer eluents of the same elution strength.
4. Adding methanol or acetonitrile will cause the pH value of the eluent to drift into the alkaline.
5. The greatest changes in selectivity usually happen around the pK_a values of the analytes, whereas the greatest robustness is achieved with a separation at a pH value that differs from the pK_a by ±2 pH units.
6. When the pH value changes from 3 to 5, with everything else remaining constant, the following parameters may change:

   - Peak height
   - Peak area
   - Retention time
   - Lifetime of the column
   - Peak symmetry
   - Plate number

7. When permanently used in an acidic (pH ca. 2) or in an alkaline (pH ca. 10) environment, silica gel will slowly but surely dissolve.
8. Selectivity permitting, one should work at a pH value of around 2.5–3.5 because many silanol groups are undissociated in this range. This reduces their interaction with polar compounds, and peaks become more symmetrical.
9. KH_2PO_4 as a buffer salt is less aggressive towards the column than $(NH_4)_2CO_3$.
10. Ionic substances elute earlier when the ionic strength (buffer strength, salt concentration) is increased.
11. The pH value of the eluent should only be measured after the addition of methanol/acetonitrile, because a considerable drift could occur if they are subsequently added to the original aqueous phase. The final pH value may not even be known.
12. Increasing the flow rate shortens the analysis time but also increases the use of solvent.

Did Peaky remember everything correctly?

1.5 General HPLC Tips

After the short break, we will continue with renewed vigor

Tip No. 56 What changes can you expect when switching from one HPLC system to another?

Problem/Question

Your pump has gone on strike, and no amount of coaxing, threatening or changing seals will make it work again. You notice that another isocratic HPLC system is sitting in the corner and is not being used. You are under pressure to finish your series of experiments, so you take your column, your eluent, your samples, etc. and try your luck with that system. What should you expect to be different from your old system, and what should remain the same – at least in theory?

Solution/Answer

Let us assume that: (a) the system has been properly flushed after the last use, (b) it is in good working order and (c) you are using the same settings as in your system, including the PC settings, such as sample rate, threshold, etc.

Consider the overview in Table 56-1.

Conclusion

When transferring gradient methods to another system, differences in mixing chamber characteristics, mixing chamber volume and dwell volume can lead to a change in retention times. In the worst case, the whole chromatogram may look completely different. By contrast, in isocratic methods, for reasons given in Table 56-1, the signal may change while the retention time could be expected to remain more or less constant. Furthermore, there may be system-specific differences, such as pressure and baseline noise.

Table 56-1.

Parameter/display	Explanations, comments and notes
This should remain the same	
Retention time	However, even if the set temperature is the same, but the conditioning method is different, there may be a change in retention time. See Tip No. 54
Selectivity (to put it simply – the distance between the tops of the peaks)	The reason being that as the chemical parameters don't change, the time in the stationary phase will remain constant
This may change	
Pressure [a]	Longer and/or thinner capillary (this hardly affects the retention time)? There should be no more than 5 to 10 bar difference
Peak area and height are decreasing	Some compounds may be irreversibly adsorbed at the surface of steel capillaries – possibly, you have been using PEEK capillaries in your system or elsewhere in the system. Or you are measuring on the UV flank, while the actual wavelength has changed compared with the original setting
The peak height is decreasing [a] → decline in resolution (to put it simply – distance between peaks at the base)	The other system has a larger dead volume
Baseline is more unsteady [a]	There is no restriction capillary behind the detector in the other system
Detection limit is higher [a]	The old UV lamp has a negative impact on the peak-to-noise ratio

a) If you experience the opposite, the comments may apply to the original system.

Tip No. 57
What changes can be expected in a chromatogram if the dead volume is larger in one isocratic system than in another?

Problem/Question

An isocratic method is going to be applied to another instrument. The system in question has a dead volume that exceeds that of the original instrument by 100 µL. What noticeable changes in the chromatogram can be expected, if any?

Solution/Answer

As you know, a dead volume has a broadening effect on peaks, which means that resolution will deteriorate, and perhaps even be insufficient where early eluting critical pairs are concerned. What else might happen? Retention factors (k values) also decrease.

Explanation

The following formula applies:

$$k = \frac{t_R - t_M}{t_M} = \frac{t_{net}}{t_M}$$

k = retention factor (formerly capacity factor k')
t_R = retention time of a component
t_M = dead time (retention time of an inert component)

Although with increasing dead volume, t_{net} (dwell time of the component on the stationary phase) remains constant, the dead time will increase because an inert component will, of course, elute later. Result: the k value decreases. The following equation will yield the same result:

$$k = K \cdot \frac{V_s}{V_m}$$

K = distribution coefficient
V_s = volume of the stationary phase
V_m = volume of the mobile phases

An increase in dead volume leads to an increase in the volume of the mobile phase. This does not affect the volume of the stationary phase or the distribution coefficient, as the chemistry does not change. This means that the k value must decrease, and more noticeably so with longer than with shorter retention times. Let me demonstrate this by the following examples:

At an increase in dead volume by 100 µL and assuming a flow rate of 1 mL min^{-1}, a delay of 0.1 min in retention time can be expected.

$$V = F \times t$$

$$t = \frac{V}{F} = \frac{0.1 \text{ mL } (100 \text{ μL})}{1 \text{ mL/min}} = 0.1 \text{ min}$$

This delay can be observed in an inert as well as in a retained compound. If, for example, we assume a dead time of 1 min in the analysis of four compounds with the retention times of 2, 4, 8 and 15 min, we would find the following new k values ($k_{100 \text{ μL}}$):

Previous

$$k = \frac{2-1}{1} = 1$$

$$k = \frac{4-1}{1} = 3$$

Now

$$k_{100 \text{ μL}} = \frac{2.1 - 1.1}{1.1} = 0.91$$

$$k_{100 \text{ μL}} = \frac{4.1 - 1.1}{1.1} = 2.73 \text{ etc.}$$

Conclusion

When transferring a method, you should not only think of the chemical parameters (stationary and mobile phase, temperature), which should remain constant, but also of the dead volume. As in the example cited, there may be minimal variation in retention time, while retention factors show a greater discrepancy. This makes the comparison of results difficult where k values and not retention times are used as criteria.

Table 57-1.

t_M	t_R	k	$k_{(100 \text{ μL})}$
1	2	1	0.91
1	4	3	2.73
1	8	7	6.36
1	15	14	12.73

Tip No. 58

Contribution of the individual modules of the system to band broadening

Problem/Question

While the overall selectivity of a system may be satisfactory, it could still be a struggle to stop the peaks from becoming wider than they should be. What reasonable steps can we take to fix this problem?

Solution/Answer

In order to answer this question, we could use the following approach.

Chromatography – and analysis in general, for that matter – is nearly always based on independent stochastic processes. These are processes in which the variance of a parameter measured in a single step of the process [e.g., the individual modules (or components) of the instrument] is independent of the variance in other steps or modules. Applied to HPLC this means, for example, that band broadening in a detector cell is independent of band broadening in the capillaries. Accordingly, the variances (variance defined as standard deviation squared) can be summed. The square root of this sum then gives the standard deviation (variance) of the total process, in this case the amount of peak broadening:

$$\sigma^2_{total} = \sigma^2_{injection} + \sigma^2_{capillaries} + \sigma^2_{column} + \sigma^2_{detector}$$

As we are interested in the practical consequences of these difficulties, we will not calculate the individual σ-values, using the appropriate formula, but look at some particular cases more closely to find out what measures can actually improve results.

In the following examples, the figures in the first row give the absolute contribution of the module in question to band broadening in μL^2, the second row gives the percentage of the contribution.

Let us initially consider the following situation:

- Column: 125×4 mm, 5 μm
- Flow: 1 mL min^{-1}
- Length of capillaries: 30 cm
- Diameter of capillaries: 0.17 mm
- Cell volume: 8 μL
- Injection volume: 10 μL

First of all we will look at the band broadening of an early peak, say after 3 min, and then of a later peak, perhaps at 8 min.

These are the figures for the peak after 3 min:

```
400   +   501   +   30.000 + 4.100
1.14%     1.43%     85.71%   11.70%
```

The peak at 8 min yields the following figures

```
400   +   501   +   83.340 + 4.100
0.45%     0.57%     94.34%   4.64%
```

What does this tell us? Firstly, the main contributor to band broadening is the column in any case. It is well worth finding a supplier who knows how to pack a better column. Secondly, the earlier the peak elutes, the stronger is the impact of the HPLC system on band broadening. This means you should inject smaller amounts, perhaps use connection tubing with a smaller internal diameter and above all think about using a smaller detector cell. Thus, if using a 4 µL cell for the earlier peak, the figures are as follows:

```
400   +   501   +   30.000 + 256
1.28%     1.61%     96.30%   0.82%
```

This means a 10% efficiency gain, as the contribution of the detector to band broadening drops from 11.70% to 0.82%, which means that the theoretical plate number of this peak is roughly the same as of a peak that elutes at about 8 min (96.30% and 94.34%) on the same column.

In other words, better use is made of the performance of the column.

For the later peak, however, reducing the cell volume would improve efficiency only by about 1%.

If we used a 3 µm column, but retained the 8 µL cell, these would be the figures for the early peak:

```
400   +   501   +   17.937 + 4.100
1.74%     2.18%     78.18%   17.87%!
```

In other words, not much is gained if we use 3 µm columns but neglect other aspects of the instrument, e.g., the cell volume. This mainly applies to the early peaks. In our case, at a cell volume of 8 µL the contribution would be near 18%, and would rise to a staggering 25.36% if the detector volume were 10 µL – a cell volume often found in detectors! I don't think there is any need to go through these calculations any further, and we can summarize.

Conclusion

1. The later the peaks elute in isocratic runs, the less important becomes the efficiency of the column (to put it simply: the quality of the packing), and other aspects of the set-up also lose their significance. Thus, in a peak that elutes after about 8 min under the conditions described above, using a 3 µm column or reducing the cell volume to 4 µL would improve efficiency by just 2%.
2. The earlier the peaks elute, the more it pays to modify the instrument to make better use of the efficiency of modern columns. A quick and easy improvement would be reducing the injection volume and using 0.12 mm capillaries. If, in addition, you would like to use a 3 µm column, investing about 1000–2000 € in a smaller cell would have long-term benefits.

Tip No. 59 — How to keep retention times constant while reducing the diameter of the column

Problem/Question

Reducing the inner diameter has two substantial advantages – the limit of detection can be lowered and the use of solvent can be reduced. Suppose you just reduce the inner diameter of the column while keeping all the other chromatographic conditions unchanged, i.e. stationary phase, eluent, temperature, etc. What flow rate should you choose in order to keep the retention times constant?

Solution/Answer

The following formula will help you calculate the flow rate you are looking for.

$$\frac{F_1}{ID_1^2} = \frac{F_2}{ID_2^2}$$

F_1 = original flow rate (mL min^{-1})
ID_1 = original inner diameter (mm)
F_2 = new flow rate (mL min^{-1})
ID_2 = new inner diameter (mm)

Setting the flow rate obtained by applying this formula will give you the same linear velocity for both experiments while keeping retention times more or less constant, as long as you do not overload the column. The formula above is applied easily enough, but I can make life even easier for you. Table 59-1 gives you the flow settings for the most frequently found column diameters.

Example: You are working with a 4.6 mm column at a flow rate of 2.5 mL min^{-1}. For a 4.0 mm column you should set the flow rate to about 1.9 mL min^{-1}, for a 3.0 mm column to 1.1 mL min^{-1} and for a 2.1 mm column to 0.5 mL min^{-1}. In all these cases, the linear velocity remains constant at 0.7 cm sec^{-1}, and all peaks elute at approximately the same time.

Conclusion

People might feel more inclined to modify their existing HPLC method if retention times could be kept constant. Table 59-1 should help in this process.

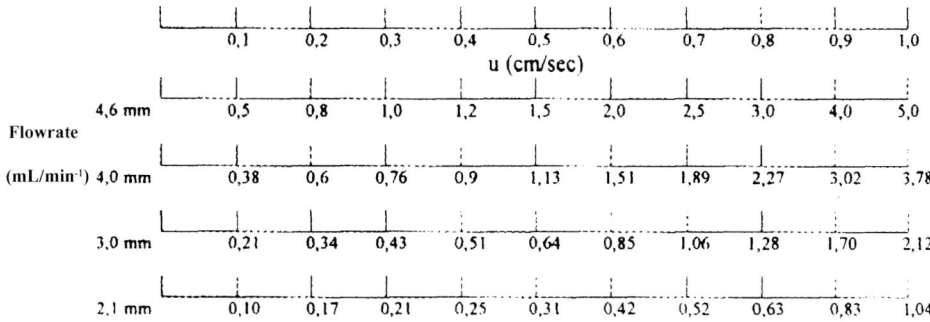

Figure 59-1. Constant linear velocity with various flow rates.

Tip No. 60 — Has 3 µm material been developed sufficiently to be used in routine separations?

Problem/Question

When I hold HPLC classes, various questions are raised that are probably on the minds of many users. One question I came across recently is: "In recent brochures the advantages of 3 µm columns over 5 µm columns have been advertised. Are they really ready to be used in routine separations?"

Solution/Answer

Yes, if certain requirements are met.

Preliminary comment: They are called 3 µm, but in most materials the actual particle diameter is 3.5 µm, the reason being that the particles have to be just large enough not to slip through the most commonly used frits and filters. Reducing the particle diameter even further would involve changing too many of the things that you are used to, and the manufacturers are very wise not to ask too much of the user at once.

Let us look briefly at the advantages!

Under the same chromatographic conditions (identical properties of stationary phase, identical packing quality, etc.) 3.5 µm material can yield the same theoretical plate number as 5 µm material, but at a higher flow rate. As a result, the analysis times are shorter. Alternatively, the same separation can be achieved at the same flow rate using a shorter column. This also shortens analysis time. The following figures will serve as an illustration. There is a rule-of-thumb formula to work out roughly the number of theoretical plates (Source: John W. Dolan and Lloyd R. Snyder, Troubleshooting LC-Systems).

$$N \approx 3000 \frac{L}{d_p}$$

L = length of column in cm
d_p = particle diameter in µm

For an ordinary 12.5 cm×5 µm column this would give

$$3000 \cdot \frac{12.5}{5}, \text{ about 7500 plates}$$

When switching to a 10 cm×3 µm column, the result will be

$$3000 \cdot \frac{10}{3.5}, \text{ about 8500 plates}$$

Even leaving aside the gain of about 1000 theoretical plates, shortening the analysis time by around 20% would be a considerable advantage. If the original column was 15 cm long, the reduction in analysis time would be by one third, as there is a linear correlation between column length and analysis time. It is therefore logical to assume that column length and flow remaining constant, 3.5 µm material will yield higher theoretical plates and thus higher peaks. This, in turn, will lower the limit of detection.

What about the downside?

1. More expensive to purchase.
2. Pressure raised by a factor of 2–3.
3. Higher demands on the instrument (extra-column band spreading).
4. If samples are contaminated, 3 µm columns are more likely to suffer.

Discussion of points 1–4:

1. This is for the individual user to decide.
2. This drawback can be at least partially compensated for by using a shorter column.
3. In gradient separations, the advantages of using 3.5 µm are only marginal, as the separations yield sharp peaks anyway. In isocratic separations, the dead volume of the instrument should not exceed about 30 µL. This means the volume of the detector cell should not be larger than 8–10 µL and the inner diameter of the capillaries no larger than 0.12 mm.
4. 3 µm columns are indeed more suitable for clean samples unless you are prepared to go to all sorts of trouble when preparing a sample.

Recommendation: Those among you who are keen on narrow peaks should perhaps make up your minds and part with those cherished 0.25 mm capillaries that were cutting-edge technology in the seventies. I know how hard it is to say farewell – don't remind me of that 18 year-old VW van sitting in my garage – but I am sure you can do it.

Conclusion

Once you have overcome the problem of the increase in pressure and found a column supplier who knows how to pack 3.5 µm material, all you need is a reasonably decent system, and you can begin some serious work on those 3.5 µm particles. By the way, even 1.8 to 2 µm particles can be used very successfully, but only if you have taken the time to optimize your instrument adequately.

Tip No. 61 — Miniaturization may be all well and good – but when does it really work and does it make sense in routine separations?

Problem/Question

For economic and analytical reasons, miniaturization is often a sensible approach. What can be achieved using an ordinary instrument, and when is it worth taking the plunge?

Solution/Answer

Preliminary comment:

I can understand why some of my colleagues are rather sceptical about miniaturization and I even share their attitude in certain circumstances, such as tightly regulated environments, samples with a high matrix content and/or a large number of components, shortage of specialists in routine analysis labs, lack of time to follow up potential errors, etc. This does not necessarily mean that in such cases miniaturization would not make sense – it only means that you have to assess your situation carefully and base your final decision on objective and verifiable facts. You should be asking questions such as these:

- How much time would it save me – not only in the analysis, but also or even more so in terms of productivity? How much work would be involved in a new registration or in a changing control procedure?
- Is there room for any further optimization in the sample preparation procedures?
- Is the selectivity sufficient to separate all the peaks clearly on a shorter column?

These are the parameters that can be minimized:
- Length and inner diameter of the column
- Particle size
- Interaction of analytes with the stationary phase

The third point should be discussed extensively; but we will just look at the various aspects concerned with the instrument.

Here are some concise hints that could be put into practice in a "real-life" lab situation:

- *Column length 60 to 100 mm, inner diameter 3 mm, particle size 5 µm*
  Preliminary condition: The dead volume of an instrument for isocratic analysis should be ≤30 to 35 µL in order to take full advantage of the miniaturization, e.g., higher peaks. For a gradient separation under the aforesaid conditions the dwell volume of the apparatus is of minor significance.
- *Column length 60 mm, inner diameter 3 mm, particle size 3 µm*
  Preliminary condition: Dead volume ≤15 to 20 µL, clean samples.

These are the advantages to be gained in both cases:

- Reduction of analysis time
- Increase in efficiency (higher peaks)
- Decrease in solvent consumption

Here are two cases in point:

1. Reducing the inner diameter from 4 to 3 mm also reduces solvent usage by about 50%. Nowadays you are very unlikely to come across a system of such poor quality that it won't work with 3 mm columns.
2. Transferring a separation from a 150 mm, 5 µm column to a 100 mm, 3 µm column (same stationary phase!) will yield the same resolution, but reduce the analysis time by around 30%!

This shows it is worth thinking about miniaturization, doesn't it?

Conclusion

A sensible isocratic system (dead volume under 50 µL) will work perfectly with the following column parameters: L 60–100 mm, ID 3 mm, d_p 5 µm, i.e., 0.12–0.17 mm connection tubing, 4–8 µL cell volume.

Switch to columns with the following parameters:

ID 2 mm and d_p 3 µm involves a lot of adjustments within the instrument. The instrument needs to be first class – dead volume around 15 µL, i.e., 0.10 mm capillaries, cell volume 2–4 µL.

Suggestions on how to proceed

First reduce the length of the column, e.g., from 125 to 100 mm, then even further to 60 mm (or you might go for 60 mm straightaway!). Once you have succeeded and are happy with the results, you should next think of reducing the inner diameter, from 4.6 to 4 mm, then to 3 mm. Once this has been successful, take a deep breath and brace yourself for a more complicated change. Optimize your instrument and then successively try out a 3 µm material and/or a 2 mm column. This is where the average user's capacity for experimenting ends. If you are allowed more leeway and have a sense of adventure, you could try the next step – capillary liquid chromatography (CLC): $\approx$ 250 µm capillaries, flow rates around 10–20 µL min^{-1}, injection volume $\leq$1 µL, cell volume 1–2 µL. This is only worth going into if you have a lot of time to experiment with (or a lot of money to buy a special CLC instrument) and nobody bothers you with other work. For more about micro and nano LC, see the contribution by Jürgen Maier-Rosenkranz in Part 2.

Tip No. 62 — Why is it that peaks appear later with a new column?

Problem/Question

The new column you ordered has just arrived. Although it has the same stationary phase and the same dimensions and you have used the same supplier, the peaks come later, however constant you keep the chromatographic conditions. Why?

Solution/Answer

Suppose the chromatographic conditions (temperature, eluent including the pH value, ionic strength, etc.) and sample preparation have indeed remained constant. Now check whether the retention time of the solvent front is also delayed. If that is the case, the flow rate has probably changed. Check your pump and look for leaks. Once you have made sure, with the help of a flow-meter or a stopwatch and a measuring cylinder, that the flow is okay, look at your column as the likely cause for the discrepancy. The density of the packing may vary from one column to the next if your supplier does not follow consistent packing procedures. You could now have a column that contains a greater amount of material than its predecessor, e.g., 0.90 g instead of 0.82 g. It is no surprise that all peaks – including the solvent front – elute later.

Let us now look at the second reason why the peaks could elute later, but when the dead time t_M remains constant. If t_M remains constant while t_R changes, so too does the retention factor k. Now this means that there is a change in the interaction between the analytes and the stationary phase, thus a chemical change has taken place. If you are certain that the chemistry that you are in control of (temperature, eluent composition, pH value, ionic strength, modifier concentration) has remained constant, the problem must lie with the supplier. Although the stationary phase carries the same label, its properties are not identical with the previous one, which shows up in the differences between various batches. However, it is also possible that you have unintentionally changed the surface of the previous column, perhaps by injecting a substance that was irreversibly sorbed, thus modifying the properties of the stationary phase and its interaction with the components, or that retention has changed because the previous column has been in use for a long time.

Conclusion

Is t_M constant or not? This is the detail that points you to the cause of your problem – be it physical changes in your system (flow), at the supplier's end (density of packing) or chemical problems either at your end (temperature, eluent) or at the supplier's end (properties of stationary phase) – unless you have tampered with the chemistry of the stationary phase.

Tip No. 63 — Column length, flow and retention times in gradient separations

Problem/Question

Background:

Reasonable interaction is the key to HPLC. In isocratic separations, for example, the relevant analytes should elute in a retention factor window of between 2 and 8. This is a sensible compromise between sufficiently strong interaction, robust conditions and acceptable retention time. By analogy, a mean retention factor $\bar{k}$ that applies to gradient separations is shown in the following equation:

$$\bar{k} = \left(\frac{t_G}{\Delta\%B}\right) \cdot \left(\frac{F}{V_m}\right) \cdot \left(\frac{100}{S}\right)$$

$\bar{k}$ = mean k value; the analyte is in the middle of the column (lengthwise)
t_G = duration of gradient (min)
F = flow (mL min^{-1})
V_m = column dead volume
$\Delta\%B$ = changes in B from the beginning to the end of the gradient
S = slope of the %B/t_G curve; for smaller molecules, S is set at about 5

An optimal $\bar{k}$ would be about 5, thus simplifying the equation, solved for the duration of the gradient, as follows:

$$\bar{k} = \left(\frac{t_G}{\Delta\%B}\right) \cdot \left(\frac{F}{V_m}\right) \cdot \left(\frac{100}{S}\right)$$

$$\frac{\bar{k}}{t_G} = \left(\frac{1}{\Delta\%B}\right) \cdot \left(\frac{F}{V_m}\right) \cdot \left(\frac{100}{S}\right)$$

$$\frac{t_G}{\bar{k}} = \frac{\Delta\%B \cdot V_m \cdot S}{100 \cdot F}$$

$$t_G = \frac{\bar{k} \cdot \Delta\%B \cdot V_m \cdot S}{100 \cdot F}$$

$$t_G \approx \frac{5 \cdot \Delta\%B \cdot V_m \cdot 5}{100 \cdot F}$$

$$t_G \approx 25 \cdot V_m \cdot \left(\frac{\Delta\%B}{100 \cdot F}\right)$$

Now let us look at the retention times in some typical combinations of column dimensions and flow.

Answer/Solution

Let us first consider a classical 250×4.6 mm column at a flow rate of 1 mL min^{-1} and a linear gradient of 20% to 80% B. The gradient duration is about 37.5 min:

$$t_G = 25 \cdot 2.5 \cdot \left(\frac{60}{100 \cdot 1}\right) = 37.5 \text{ min}$$

Comment 1:

It turns out that in contrast to the isocratic mode, the column length does not usually matter so much in gradient separation. The sample is effectively concentrated onto the head of the column, and while the proportion of methanol or acetonitrile in the eluent is increasing, the individual components of the sample leave the stationary phase one by one. The column length and thus the number of theoretical plates is less crucial here, since the peaks are sharp and narrow anyway.

On a 125×4 mm column under the same conditions, the duration of the gradient is around 18.75 min:

$$t_G = 25 \cdot 1.25 \cdot \left(\frac{60}{100 \cdot 1}\right) = 18.75 \text{ min}$$

Comment 2:

We have already discussed the gradient volume (gradient duration×flow) – the larger the better (e.g., increasing the flow while keeping the gradient duration constant) or faster (increasing the flow while the gradient time becomes shorter) the separation will be. For the column just mentioned (125×4 mm), a gradient run at a flow rate of 2 mL min^{-1} will take 9.38 min – and the resolution is the same as in the first two cases!

$$t_G = 25 \cdot 1.25 \cdot \left(\frac{60}{100 \cdot 2}\right) = 9.375 \text{ min}$$

In fairly straightforward separations (the number of peaks is less than 10) you should always think about using shorter columns. On a 75×4 mm column and a flow rate of 2 mL min^{-1} the gradient duration is around 5.63 min:

$$t_G = 25 \cdot 0.75 \cdot \left(\frac{60}{100 \cdot 2}\right) = 5.625 \text{ min}$$

If there are only 4 to 5 peaks to separate, a column of 20–30 mm will be sufficient in any case. The chromatogram in Figure 63-1 was produced during an HPLC class given by Daniel Stauffer, Roche, Basel, using a 25 mm column and an ancient instrument. There is, of course, plenty of scope for further optimization.

Figure 63-1. Separation of alkyl benzenes using a linear water/methanol gradient on a 25 × 4 mm column at various flow rates.

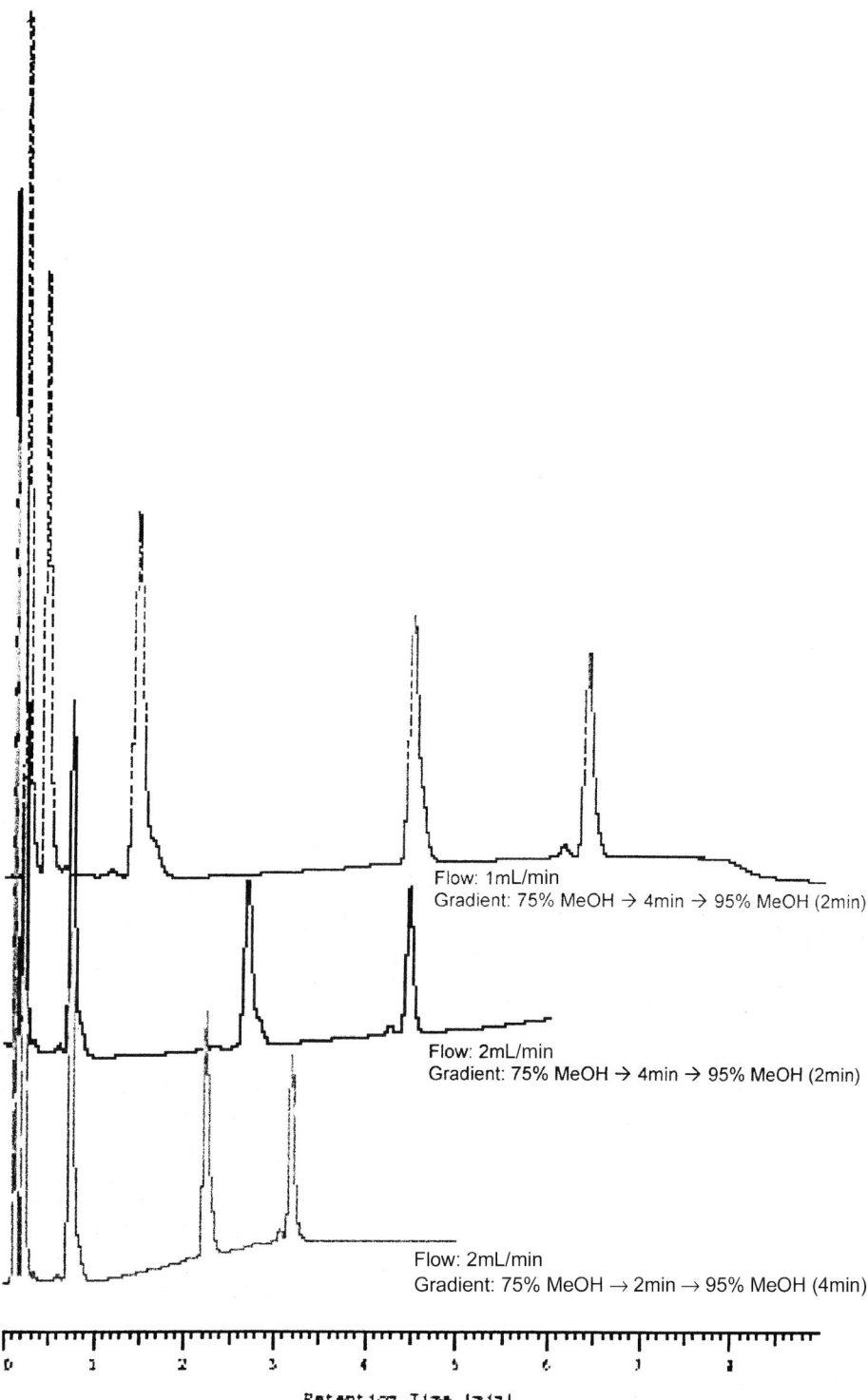

Conclusion

If there is not a good reason to use one, such as a badly contaminated sample or a problem matrix, long columns are not a good idea for gradient separations – they are simply superfluous. If you expect fewer than 10–15 peaks, even 125 mm could seem rather long.

Once you have sorted out the separation set-up you should increase the flow. You will be pleased with the result – either the separation will be even better or you will achieve approximately the same resolution in a shorter period of time. The only disadvantage is that the peak areas will become smaller, but this is the subject of another discussion.

Tip No. 64 — Column dimensions and gradient separations

Problem/Question

I know we have already discussed column dimensions for gradient separations in Tip No. 63, but after several discussions I have had with colleagues out there in the labs I feel very strongly that the subject has not yet been exhausted.

How often do I come across the following situation in gradient separations: 125×4 mm column, flow rate 1 mL min^{-1}, gradient duration 20 min. This can be easily changed. If the flow rate were set at 1.5 mL min^{-1} the same resolution would be achieved after 13–14 min. In both cases the gradient volume remains roughly the same: 1 mL min^{-1} × 20 min = 20 mL and 1.5 mL min^{-1} × 13 min = 19.5 mL. No doubt, setting the desired gradient volume (and thus the resolution) by adjusting the flow rather than the gradient duration (analysis time decreases!) is a very elegant method. However, this time we are talking about the volume or the dimensions of the column. How important are they in gradient separations?

Solution/Answer

Not very important, really! Normally, a simple separation will run on a 125×4 mm as well as on a 50 or a 75×4 mm column. This means in practical terms that if you expect only 7–10 peaks you should consider abandoning the classic 125×4 mm in favour of a 50 or 75×4 or 4.6 mm. That will save you time.

To give you food for thought I would like to tell you about a case where even mini-columns or even pre-columns were used successfully in an ordinary system. In a company with which we cooperate we ran a quick experiment – trying to transfer a simple separation of five peaks from a 150×4.6 mm to a 10×2 mm, 2 µm column without changing anything else in the set-up. It was an immediate success, and the retention time was around 4 min. See Figure 64-1. Of course, there is still plenty of room for improvement, but in this case, we only wanted to make our point.

Conclusion

During a gradient, the sample is enriched at the head of the column and then eluted from the column by raising the elution strength of the eluent. In relatively straightforward separations where no cumbersome matrices have to be dealt with, the volume of the column is usually of minor significance, and the optimization should be achieved through adjustment of the gradient volume. In straightforward separations it is worth thinking twice about using the classical arrangement of 125×4 mm, C_{18}, 5 µm, 1 mL min^{-1}. Remember that small is beautiful – at least in HPLC!

For a detailed discussion on gradient optimization see [2].

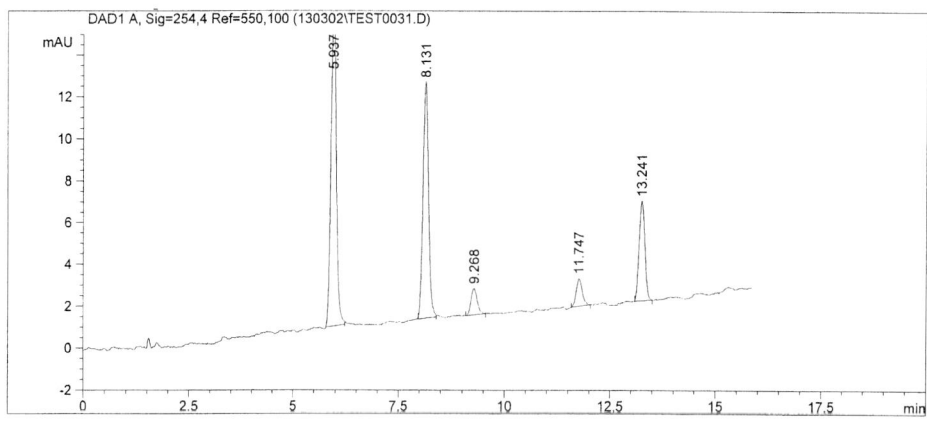

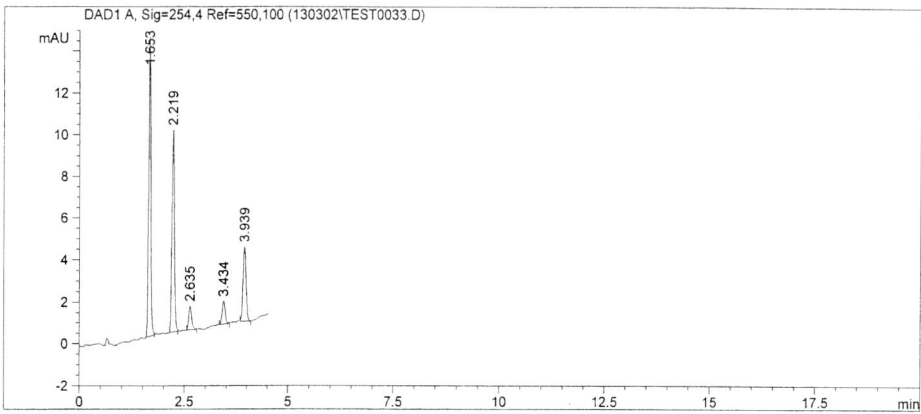

Figure 64-1. Top: Separation of 5 components on a 150 × 4 mm column, analysis time 12 min. Bottom: same separation on a 10 × 2 mm column, analysis time 4 min.

Tip No. 65 — What is the difference between dead time and dead volume on the one hand and selectivity and resolution on the other?

Problem/Question

Again and again I find that terminology is not always used in the same way in HPLC lab-speak. As this must result in confusion, I would like to give clear-cut definitions to at least some of the terms.

Solution/Answer

Dead time t_0 recently also referred to as t_M (retention time of an unretained peak, solvent peak, front, breakthrough time) is the time all analytes remain in the eluent, i.e., it is the same for all compounds. In other words, it is the time that an inert compound (a compound that can diffuse into all the pores of the packing material but does not interact with the stationary phase) remains in the eluent. In an RP system, uracil, thiourea or potassium nitrate could be used as markers. Dead time can also be calculated using the following rule of thumb:

$$t_M \approx 0.1 \; \frac{L}{F} \; \text{(cm) for 4.6 mm columns} \; \text{(mL min}^{-1}\text{)}$$

$$t_M \approx 0.08 \; \frac{L}{F} \; \text{(cm) for 4.0 mm columns} \; \text{(mL min}^{-1}\text{)}$$

Example: For a 125×4 mm column at a flow rate of 1 mL min^{-1} the dead time t_M could be expected to lie around.

$$t_M \approx 0.08 \cdot \frac{12.5}{1} \approx 1 \; \text{min}$$

Note:

Dead time depends on flow, packing density and dimensions of the column. It is independent of the stationary phase, the eluent or the temperature. *Dead volume* is the volume of the isocratic instrument excluding the column, i.e., the volume of capillaries, fittings, etc., between the injector and the detector, including the cell. It is better to call this the extra-column volume, to differentiate it clearly from the column dead time defined above. Depending on the manufacturer, the extra-column volume may lie between 20 and 120 µL!

So just remember: Dead *time* always includes the column
Dead *volume* always excludes the column

The word dead volume is also sometimes used in the context of column packing and refers to the hollow spaces that can develop in the packing over time. It has nothing to do with the dead volume (i.e., the extra-column volume) of the instrument, but it will kill the peak shapes.

Selectivity describes the ability of a chromatographic system to separate components. The chromatographic system is a combination of stationary phase and eluent including additives and temperature. The measure for selectivity is the separation factor a (formerly known as selectivity factor).

$$a = \frac{\text{time component A remains on stationary phase}}{\text{time component B remains on stationary phase}}$$

The prerequisite of a separation is that $a>1$.

Resolution – to put it simply – is the distance between the peaks at the base. Naturally, this is dependent on selectivity, but also on the degree of interaction (capacity factor) and efficiency (= separation performance measured in theoretical plates) of a column. Figure 65-1 illustrates the difference between selectivity and resolution in a diagram.

The selectivity is the same in (a) and (b) (same distance from peak tip to peak tip) because the substances remain on the stationary phase for the same amount of time. The resolution, however, is poorer in (b) because the efficiency (e.g., due to poorer quality packing or a larger dead volume because of wider capillaries) is lower in (b) than in (a).

Conclusion

I hate anyone being pedantic and red tape as much as you do, but being clear about terminology makes life much easier for everybody who deals with HPLC. The ghost of Halloween will love us for explaining clearly all the different dead things in an HPLC instrument.

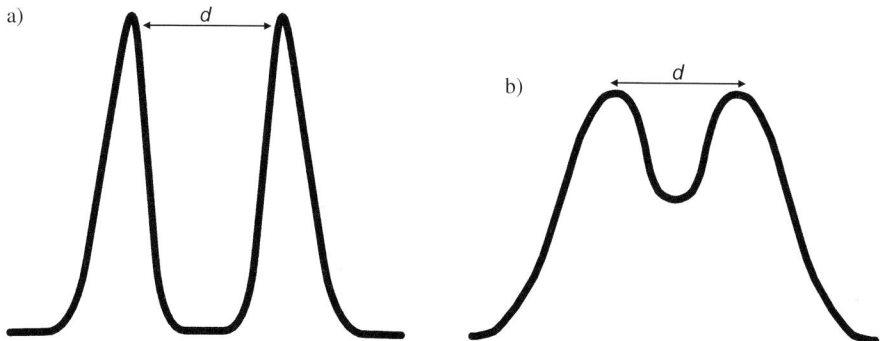

Figure 65-1. (a) Good selectivity; good resolution. (b) Good selectivity; poor resolution (due to inefficiency of the column = poor separation performance).

Tip No. 66

Troublesome small peaks

Problem/Question

"Small is beautiful" may apply to children, cats, dogs, ladybirds and the like. It is also true for the particles in a chromatography column. We might still agree it is true for a house with a garden, but when it comes to cars not everybody would subscribe to this principle, and we must definitely draw the line where peaks are concerned. Tip 38/1 explains in great detail how peaks can be made taller. What is it that makes small peaks so troublesome?

Solution/Answer

One thing is certain – small peaks that have not been recognised can compromise analysis results when considering the purity of samples. What about small peaks that have clearly been identified – could they lead to problems, too? Yes, and they mainly fall into three categories:

- Major errors in the evaluation of the peak area. Remedy: quantitation by peak height.
- Small peaks often stand for small amounts, which could easily be adsorbed by steel and glass surfaces.
  Verification: Compare the total area of all peaks (100% method) in an ordinary injection with the peak area that results from injecting the sample solution directly into the injector cell. If the area remains roughly the same in both cases, there is no irreversible adsorption of sample constituents anywhere in the system. If there is a discrepancy you should play it safe and passivate the system or de-activate the surfaces that come into contact with the sample, e.g. treat glass surfaces with trimethyl silane (TMS) and optimize the sample solvent.
- When voluntary or involuntary changes occur within a chromatographic system (pH value, temperature, etc.) the small peaks tend to wander more easily than the tall ones. This may result in reversals of the elution order and a game of hide-and-seek with the major peak.
  Remedy: All trace analysis methods require a high degree of robustness (for more about robustness see Tip No. 55).

For verification purposes, vary the standard conditions as follows:

- ±5% ACN
- ±5 °C
- ±0.05 pH
  and compare retention times and areas.

Conclusion

If you are worried about an inadequate detection limit, there are several simple solutions at hand that will result in sharp peaks. For reproducible retention times and peak areas, robust chromatographic conditions are indispensable.

Tip No. 67 Lowering the detection limit by optimizing the injection

Problem/Question

Nowhere is the improvement of peak shapes as crucial as in trace analysis (purity, stability and toxicity tests, cleaning validation, environmental analysis). If the amount injected and thus the peak area are kept constant and there is a discrepancy in peak height, then there must be a minor error in quantification. There are several time-honoured methods that will result in narrower or higher peaks, e.g., thinner or shorter columns, smaller particles, lowering the dead volume in the instrument, optimization of detection, improving the kinetics (temperature, modifier), etc. Here we will discuss some simple tricks used in the injection process.

Solution/Answer

There are two approaches:

1. Preventing dilution of the sample zone
2. Concentrating the sample onto the head of the column (enrichment)

Discussion of 1:
Laminar flow causes a dilution of the sample zone. This must be prevented.

- Inject 10 to 20 µL of air together with your sample (take up the air first, then the sample) in order to create an air cushion that prevents the dilution of the sample zone travelling from the injector to the column. See Figure 67-1. Incidentally, there are injectors that produce such air segments automatically.
- Inject some guanidine, urea or 2% glycerol together with your sample. These substances act as highly viscous stoppers that serve the same purpose as the air bubbles. Since these substances elute with the solvent front, they won't get in the way. See Figure 67-2.
- Does the loop volume match the injection volume? Did you connect the injection valve correctly? Please check!

Discussion of 2:
You can try and concentrate the sample onto the column head. This will succeed if the sample solvent is weaker (from an elution strength perspective) than the eluent.

Here are three typical cases:
Neutral samples, RP system: The sample solvent should be more polar than the eluent, e.g., eluent 50/50 acetonitrile/water, sample solvent 25/75 acetonitrile/water.

Ionic substances: Increase the ionic/buffer strength – and thus the polarity – of the sample solvent, i.e., take a fluid containing more salt than the eluent to dilute the sample, e.g., eluent 20 mmol phosphate buffer, 50 mmol buffer for the sample dilution.

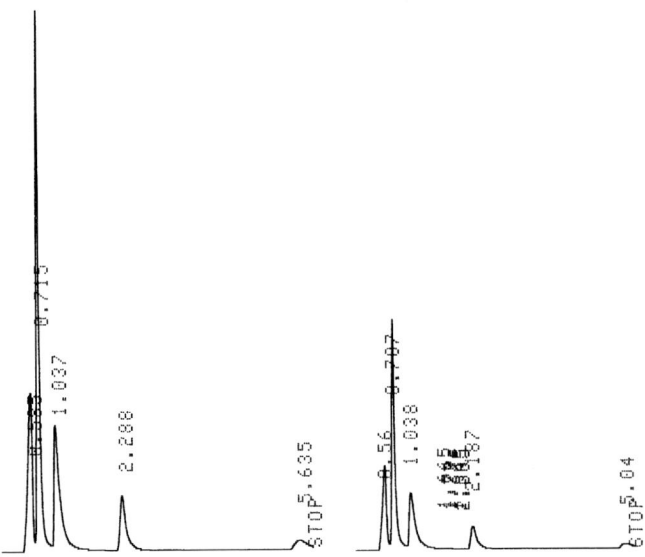

Figure 67-1. Improving the peak shape at the start of a chromatogram using an air cushion (right panel). Comments see text.

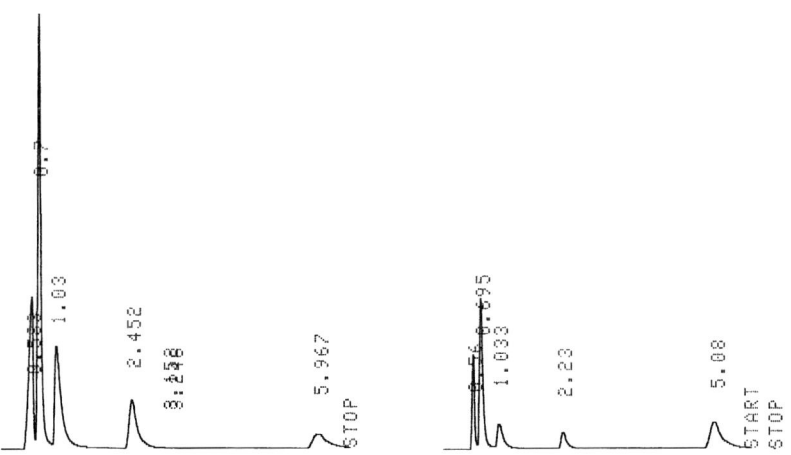

Figure 67-2. Improving the peak shape by increasing the viscosity of the sample solution (right panel). Comments see text.

Organic bases: For example if working in an acidic medium, dilute the solvent in a weakly alkaline solvent. The bases are now deprotonated and can be more easily retained at the head of the column.

Conclusion

If you work in trace analysis, it is worth thinking about optimizing your injection technique.

Tip No. 68 — Setting the parameters of an HPLC instrument

Problem/Question

You are probably aware of the importance of the correct setting of parameters in an analytic tool, and yet I come across so many systems that are not set at their optimum for the particular purpose. What a waste of opportunity, as there is such scope for achieving sufficient resolution without having to change the column or the eluent. What settings are we talking about?

Solution/Answer

I think we can skip the obvious parameters such as flow, temperature or injection volume. Table 68-1 gives a brief overview of "secondary" parameters.

More about integrator/PC settings and their impact in this context can be found in the section by Hans-Joachim Kuss in Part 2.

Each of the points mentioned in the table can be illustrated by an example. Let us look at the time constant. See Figures 68-1 and 68-2.

If we work with a time constant of 1.1 s (Figure 68-1) we lose out unnecessarily on resolution and detection limit, compared with working with a time constant of 0.1 s under otherwise equal conditions. This means that using a high time constant prevents us from fully exploiting the separation capacity of modern materials.

Figure 68-2 shows three chromatograms – top, at a time constant of 4 s, middle at 2 s and lower at 0.1 s. In the top panel, a hump can be distinguished at 1.6 min, while at 1.3 min there could be something, but this is no more than a suspicion. The graph in the middle clearly shows two peaks on the flank of the large peak, and in the diagram below a peak integration is actually possible without any changes to the chromatographic conditions.

At the top of Figure 68-3, a time constant of 4 s yields three peaks – one of them sitting on the flank of another peak. Once the time constant is reduced to 0.1 s, a further component can be clearly distinguished from the main peak. Also note the narrow shape of the first peak. Low time constants can lower the detection limit considerably.

Comment:

If a change of the time constant does not result in a noticeable change in peak shape, the cause might well lie in the output of your detector. Check if it is better to use the 10 or 100 mV output. Or you manipulate at the "bunching factor" or "bunching rate". See also the contribution from Hans-Joachim Kuss in Part 2.

Conclusion

Before you go through the trouble of changing a column or eluent you should optimise the instrument settings, which can be done in a few seconds at the most!

Table 68-1.

Parameter	Effect	Comment
Pump: high (low) pressure limitation	Automatically switches on or off when a given pressure value is reached	This is often set too low, depriving you of the opportunity to work with higher flow rates and cut down on analysis time. Ordinary columns can take pressures up to 300–350 bar with hardly any loss in efficiency
Injector: aspiration rate	Speed at which the sample solution is taken up by the syringe	If the speed is too high for a viscous solution, air bubbles will form and affect the reproducibility of the injection. Also when using water as the sample solvent choose a slow aspiration rate!
Detector 1. Time constant, also known as response time, or rise time	This is the time a detector needs to register 66% of a signal, in other words the response time of the detector: how long does the detector take to respond to the analyte?	A small time constant (e.g. 0.1 s) results in narrow peaks while a large one produces wide peaks (e.g. 2 s)
2. Slit width in DAD	Controlling the amount of light that falls into the detector cell	A narrow slit gives good optical resolution (good UV-spectrum) but affects sensitivity (detection limit goes up), while a wide slit keeps the detection limit low at the expense of good optical resolution
PC: sample rate	The number of data recorded per second	Same effect as with the time constant, e.g., at a sample rate of two data points per second narrow early-eluting peaks can be completely obliterated unless they elute in a broad band

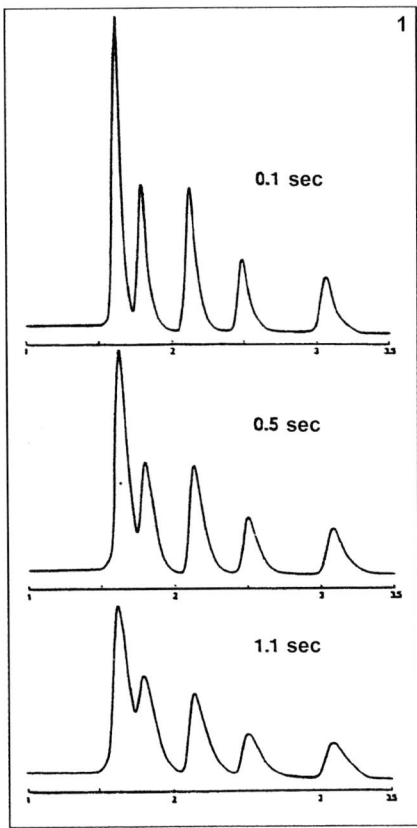

Figure 68-1.

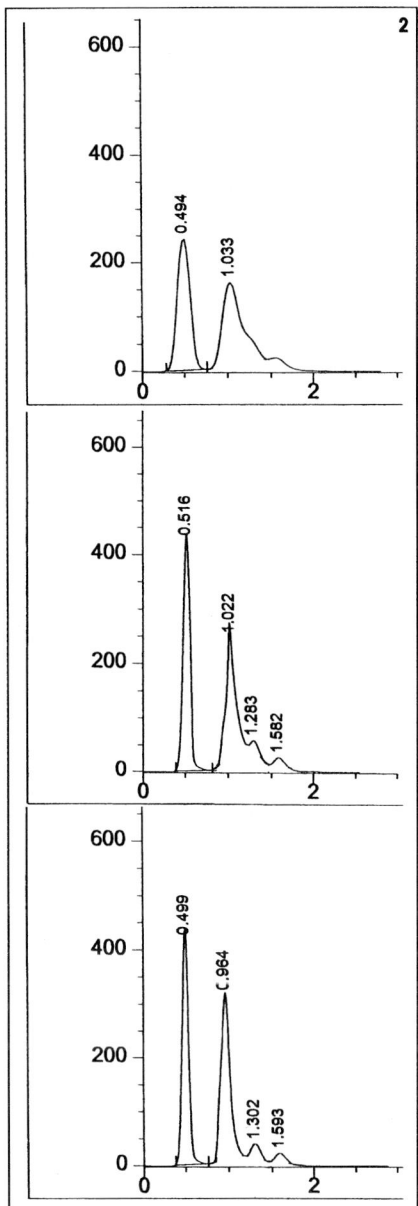

Figure 68-2.

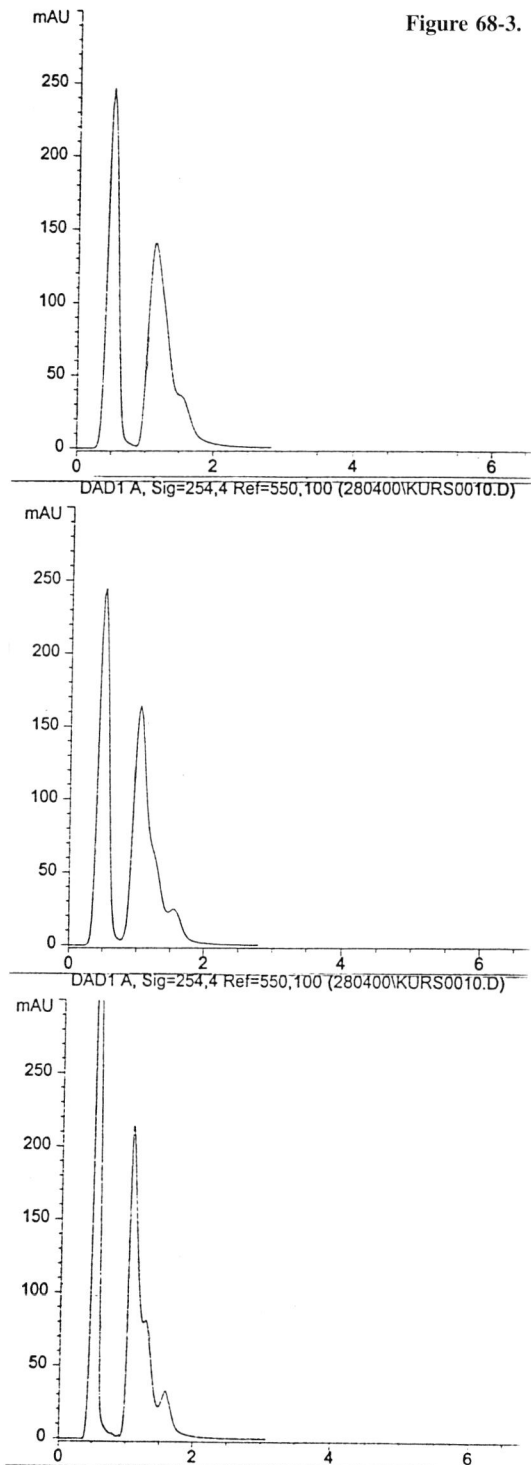

Figure 68-3.

Tip No. 69

The right wavelength – old hat to some, a revelation to others

Problem/Question

For quite a while, I was not at all sure whether I should write an HPLC tip on such trivia, but after I spent half a day on a wild-goose chase for lost peaks and ended up swearing in Greek, standard German and the Saarbruecken dialect, I decided something had to be done about this. I urgently advise you before you begin optimizing a separation or look for errors, to check whether your wavelength is set at the optimum or at least correctly. Well, you may think I am exaggerating – until you get yourself into the same trouble.

Solution/Answer

Well, maybe you are more meticulous or smarter than most of the rest of us, but here are three real-life examples of this problem:

Analysis of active ingredients (Figure 69-1):

There was a lively discussion in a lab on how best to integrate the peak (a contamination) sitting on the flank of the first peak (left panel, 254 nm). An injection at 300 nm yielded a chromatogram as shown on the right. Need I say more, especially as, given the correct scale, the peaks can be plotted just as "wonderfully" as on the left?

Pharmaceutical analysis (Figure 69-2):

To integrate the peak on the flank at 1.66 (right panel, 220 nm) is a brave decision, to say the least. At 280 nm the integration is quite unproblematic, see the left panel.

Environmental analysis (Figure 69-3):

When monitoring the direct discharge of waste water it is sufficient to monitor the areas of certain lead substances – peaks Nos. 6, 7 and 45 in our example. At 254 nm, peak No. 45 cannot be properly integrated (top panel). To improve the chromatographic resolution would probably be a complicated procedure. At 280 nm, however, quantification does not pose a problem (bottom panel).

Conclusion

Before you do anything else, optimize your instrument settings. This should become second nature to you. Here we discussed the detector settings. See also Tip No. 68 for some things that you can do with respect to the injection.

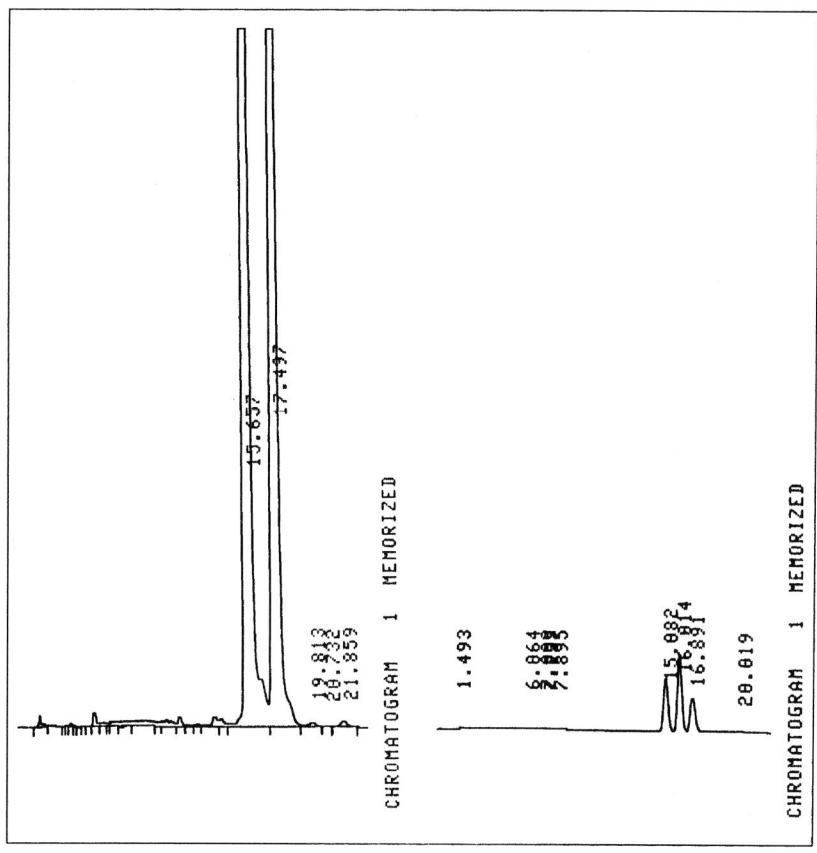

1: Active ingredient

Figure 69-1. Impact of wavelength on the separation of two active substances plus contaminations. Comments see text.

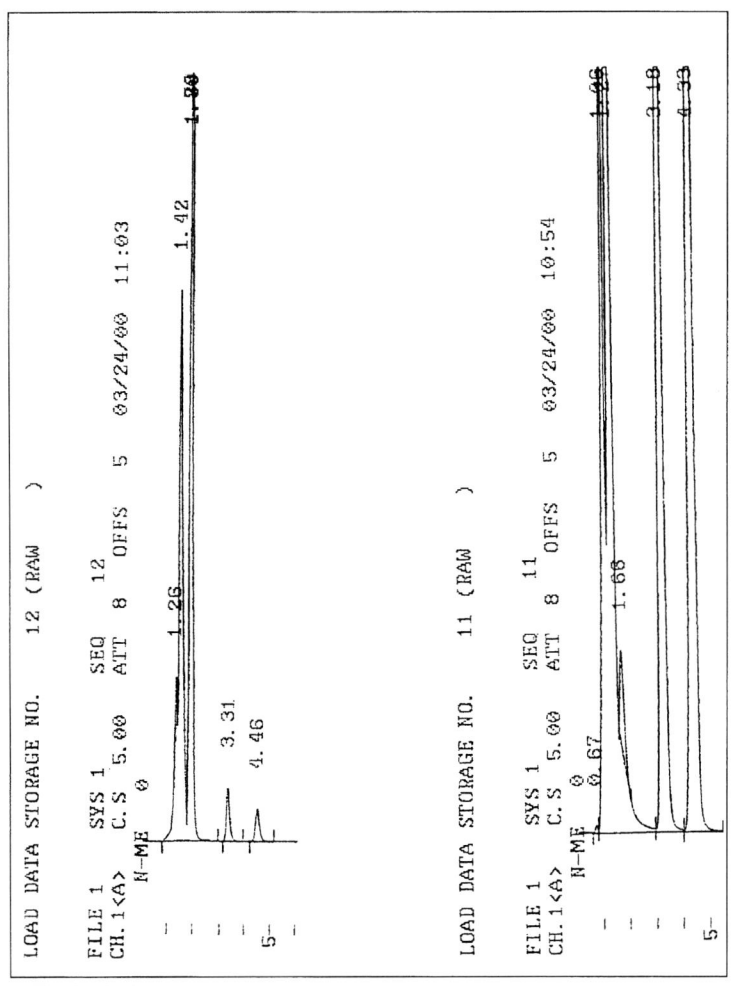

2: **Typical pharmaceutical analysis**

Figure 69-2. Impact of wavelength when separating several compounds in a tablet. Comments see text.

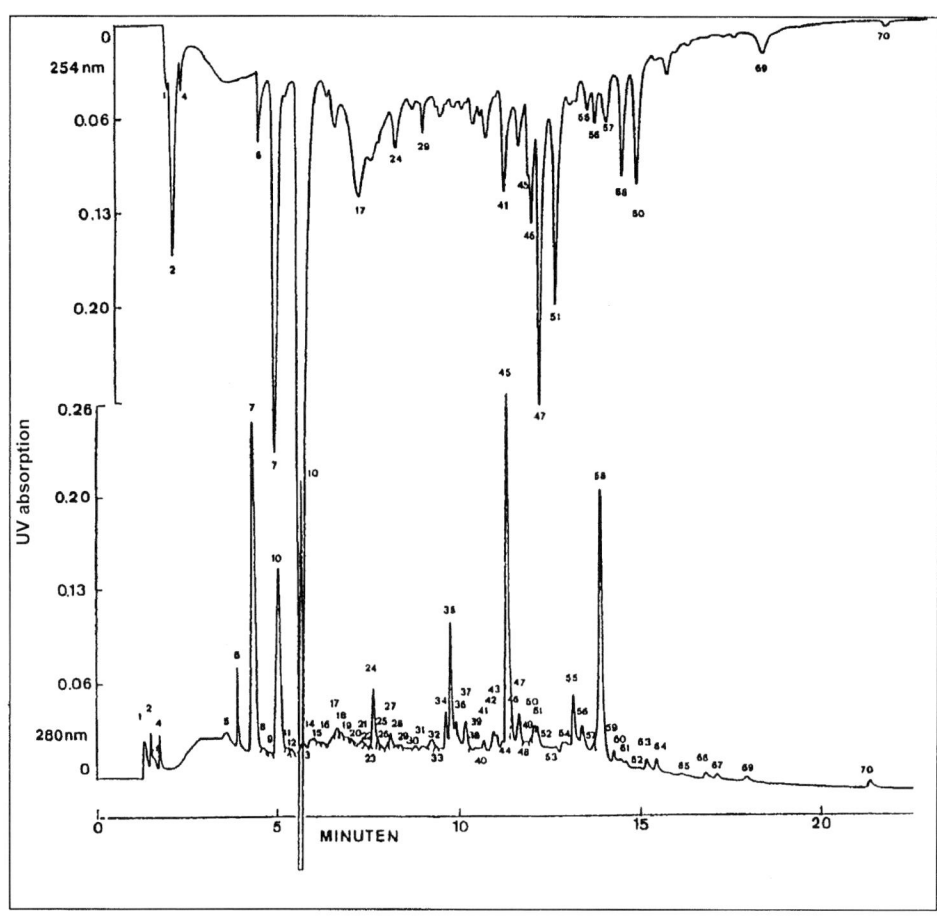

3: Environmental analysis

Figure 69-3. Impact of wavelength when separating several compounds in a waste water sample. Comments see text.

Tip No. 70 — Characteristics of refraction, fluorescence and conductivity detectors

Problem/Question

You have been working with a UV detector for years, but now you are moving into a new area, and your boss wants to buy a new detector that can detect non-UV-active substances in order to increase the specificity of the detection system. There is not enough money for LC-MS coupling, so the decision is between a refractive index detector, a fluorescence detector and – in the case of ion detection – a conductivity detector. As far as you are concerned, you only knew that such detectors existed. In what way do they differ from your dear old UV detector?

Solution/Answer

One thing that can be said of all three of them – they are rather sluggish and hate quick changes, e.g., in eluent composition, temperature or flow rate. Once they are up and running they run well, but you should not overburden them with 2 or 3 different applications per day.

Table 70-1 gives you an overview of the specific properties of these three types of detector in comparison with a variable UV detector.

Table 70-1.

	Refractive index detector	Fluorescence detector	Conductivity detector
Longer equilibration time	Yes	Yes	(Yes)
Constant temperature important	Yes	Yes	Yes
Pulsation-free flow required	Yes	Yes	Yes
Pressure stability of the cell a problem	Yes	No	No
Eluents must be thoroughly degassed	Yes	Yes	No
Small linear range	No	Yes	No
Sensitivity	Lower	Higher	Lower
Specificity	Lower	Higher	Higher
Universal use	Yes	No	No
Robustness	Less	Less	Less
Suitable for gradients	No	Yes [a]	Not really

a) Frequent problems with drift and baseline fluctuation.

Conclusion

The binary ("yes/no") statements in the table are just a rough guide. In trace analysis, for example, consistency in flow is crucial even when a UV detector is used, and a conductivity detector can detect minute ion concentrations only if the eluent has been thoroughly degassed.

Some of these detectors are very specific, which makes them indispensable for certain applications (sugars, inorganic ions, PAHs, etc.). As long as you accept their idiosyncrasies and let them do their job (a very therapeutic exercise for interfering busybodies!) they will perform wonderfully for you.

Tip No. 71 — Does it always have to be HPLC?

Some of you may ask "What a strange question to ask – and in an HPLC textbook of all places?" Well, let us have a closer look...

Problem/Question

You are determining the amount of an active ingredient, for example, from a tablet or after dissolution. Or it could be that you have to check whether the container is absolutely clean before it is used for the next production batch (cleaning validation). You may end up with a single peak. Your column is short, say 50 mm, and the chromatogram takes only 6 min. If everything goes well and you are happy then there is no need to change anything. Just skip the rest of this page and go on to the next tip.

However, at seminars and in labs I keep hearing complaints about such cases, for example:

- We have a memory effect. The area keeps decreasing. The retention time is not constant. It seems that the problem lies within the column.
- A retention time of 6 min is far too long. We need a greater throughput for the dissolution test.

What can be done in these cases?

Solution/Answer

Just leave out the column!

Since we are dealing with only one peak, we do not really need a column because there is no need for separation. What matters is the peak area, not the chromatography. There is nothing for it but to replace the column with a capillary.

Use a PEEK (polyether ether ketone) restriction capillary instead of a column, maybe 1–2 m long. Set the flow to 0.3–0.5 mL min^{-1}. This low flow rate requires a smaller injection volume (remember, flow × peak area = constant). The restricting capillary should have 10–20 knots, see Figures 71-1 and 71-2 in order to avoid those eddies and laminar flow that tend to broaden the peaks.

Note: Diffusion in a laminar flow is slow, with turbulent flow it is quicker. With turbulent currents, the middle velocity amounts to 80%, with laminar flow approximately 50% of the maximum velocity (velocity vector). The length of the capillary and the injection volume must be carefully matched.

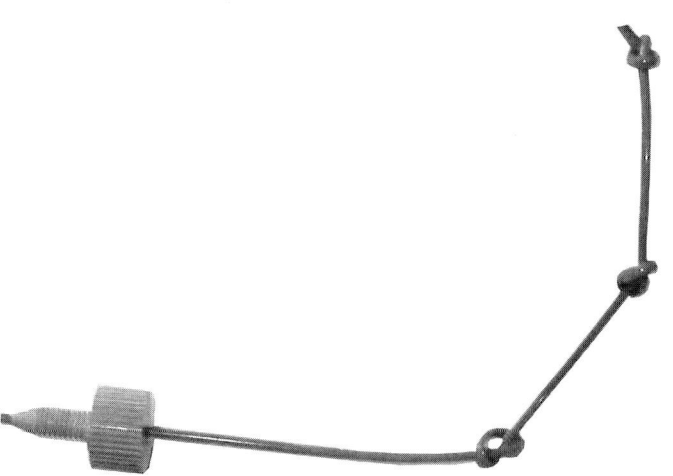

Figure 71-1. Capillary with knots. Comments see text (Source: Daniel Stauffer, Roche, Basel, Switzerland).

Conclusion

Sometimes, analysis is about identification rather than separation, in which case you should have the courage to leave out the column. You can thus avoid a major source of errors and dramatically increase the throughput in your routine analysis. Using your real standards (analyte plus matrix, auxiliary substances, salicylic acid or prednisolone in dissolution tests), you can check whether the changes you made in your instrument set-up had a positive or negative effect on variance, robustness, etc.

If you are still not convinced that you can do without a column, just use a 10 mm or 20 mm pre-column as your separation column – that's all you really need.

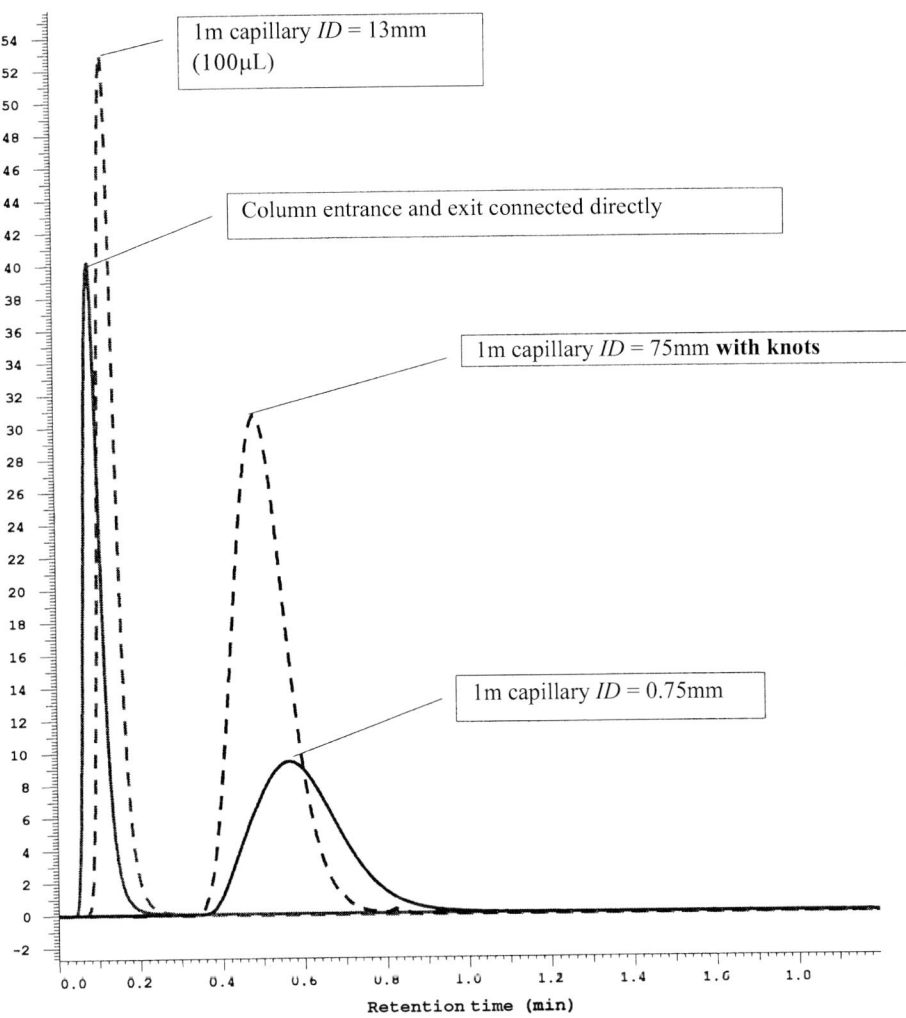

Figure 71-2. Peak shape after the substance has passed through an ordinary capillary (retention time 0.55 min) and improved peak shape as a result of the substance passing through the same capillary with knots. (Source: Daniel Stauffer, Roche, Basel, Switzerland).

Tip No. 72 Methanol versus acetonitrile

Problem/Question

Methanol and acetonitrile are without any doubt the most important inorganic solvents in RP chromatography. We have already discussed their pros and cons in Tip No. 06/1. Let us just recapitulate briefly.

Acetonitrile has a lower viscosity, which reduces the pressure slightly and usually results in better peak shapes. It helps prolong the lifetime of the columns. Moreover, its lower UV absorption in the lower wavelength range can be an advantage.

Methanol is cheaper and supposedly less damaging to human health. As methanol is polar, when using buffers, the risk of salts precipitating in narrow capillaries is lower. Methanol is less prone to contamination.

What are the differences from a purely chromatographic perspective?

Solution/Answer

Methanol, being more polar than acetonitrile, can act as proton donor as well as acceptor and form methanolates. Acetonitrile, by contrast, is aprotic. We can thus draw the following conclusions, which can be confirmed by numerous experiments:

1. Methanol encourages polar/ionic interaction. As in RP chromatography, it is the polar rather than the non-polar interactions that matter for selectivity [2]. In eluents with the same elution strength, those containing methanol usually achieve better selectivity than those containing acetonitrile – at the price of longer retention times and more asymmetry in the peaks, as will be demonstrated in two examples. Figure 72-1 shows the injection of uracil, pyridine, benzylamine and phenol, in an acidic methanol/phosphate buffer on the left, in an acidic acetonitrile/phosphate buffer on the right. These fairly unsuitable eluents (strong bases in an acidic medium) have been chosen on purpose to test the selectivity of methanol and acetonitrile for polar analytes in difficult situations. In methanol the bases were at least partly separated, which does not happen in acetonitrile. Figure 72-2 shows the same separation in the neutral, on the left in methanol, on the right in acetonitrile. Again, the better selectivity of methanol is striking. Not only are polar contaminants almost completely separated from uracil while they are barely visible in acetonitrile (see arrow), but in phenol they can also be separated from benzylamine, which does not happen in acetonitrile. We can thus conclude that methanol achieves better selectivity while acetonitrile achieves a better peak symmetry. This can be observed for many categories of substances.
2. In difficult gradient separations, you often have to begin with a high aqueous proportion in eluent A, e.g., 90 or even 95%. There is a risk that these highly aqueous eluents will not wet the surface of a hydrophobic phase. With the organic proportion in the eluent increasing, the surface will gradually become more wettable – the less polar the organic solvent, the quicker the wetting process so acetonitrile has the edge here.

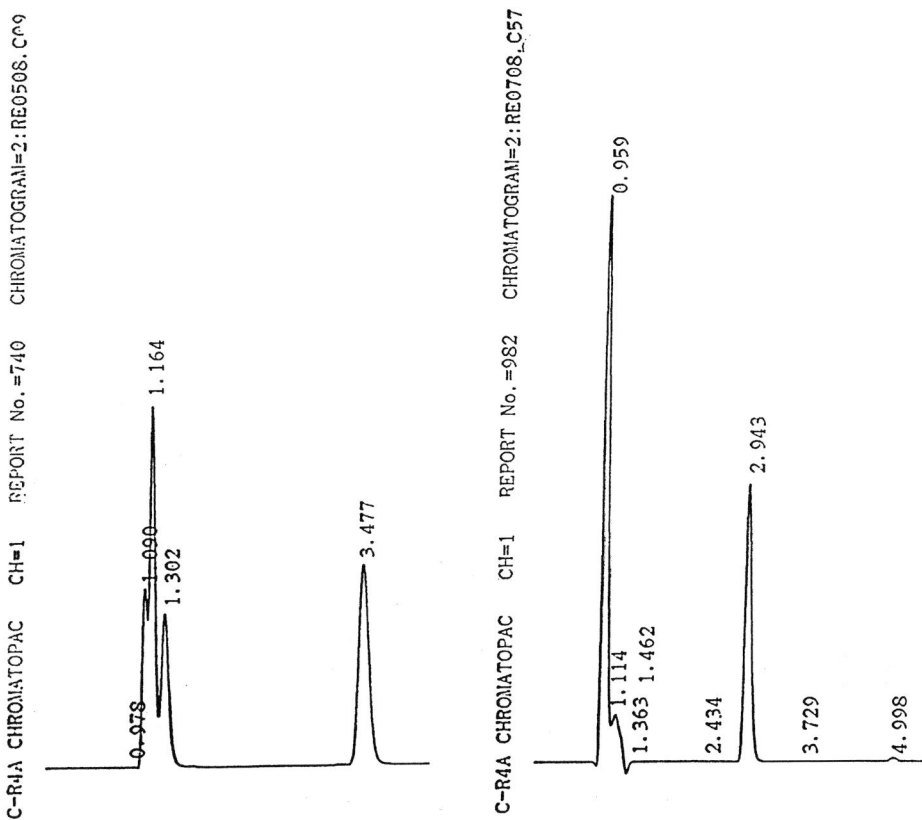

Figure 72-1. Selectivity in acid phosphate buffer/methanol (left) and in acid phosphate buffer/acetonitrile (right). Comments see text.

Conclusion

1. Eluents containing methanol can enhance the ability of phases for polar interaction. In difficult separations – and only then! – preference should be given to methanol. The resulting broadening of peaks is unavoidable.
   Remember to use methanol for the separation of stronger acids, strongly polar metabolites and isomers. Choose acetonitrile for neutral molecules and organic bases provided the selectivity is sufficient (see also the tips in the section on phases).
2. In extreme gradient runs on hydrophobic RP phases it is better to choose acetonitrile. Thus, the phase surface remains wetted over a wider range of polarity than if using methanol, and acetonitrile ensures a better reproducibility of retention time in the early peaks than methanol would.

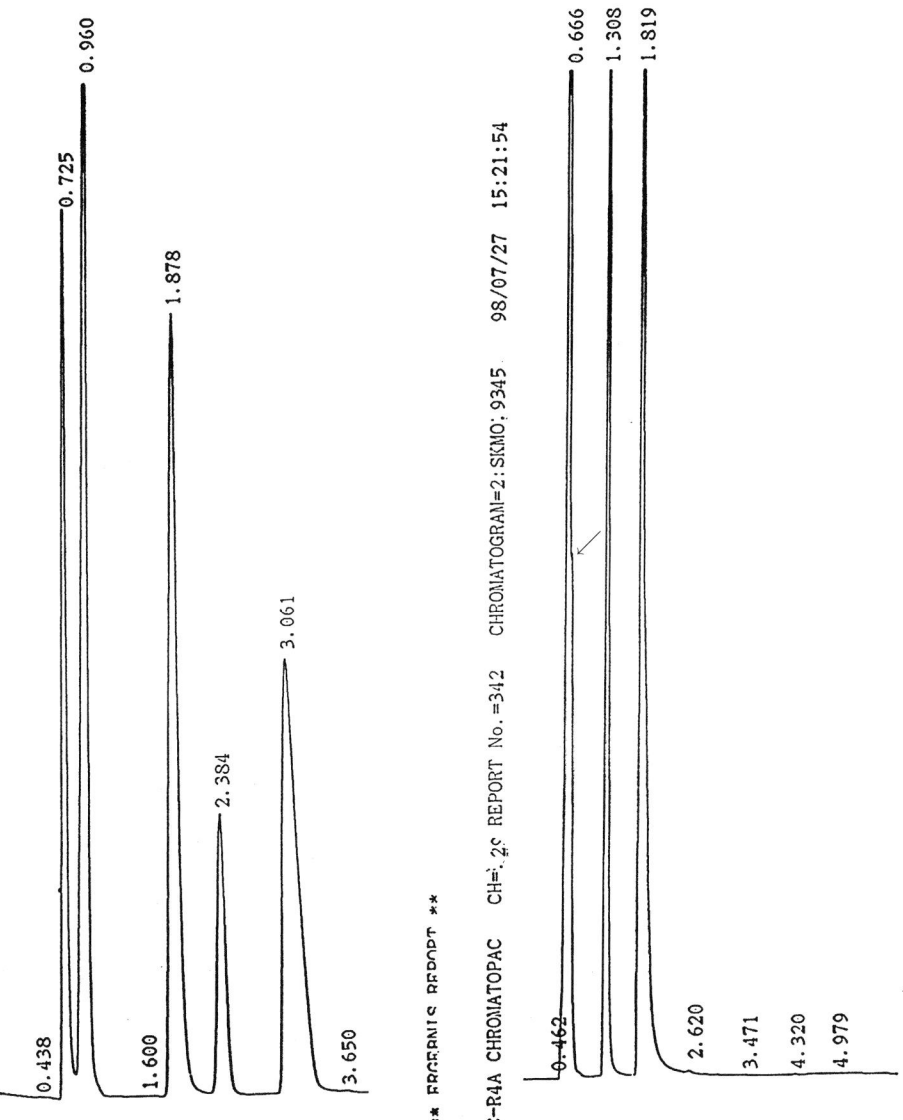

Figure 72-2. Selectivity in neutral water/methanol eluent (left) and in neutral water/acetonitrile eluent (right). Comments see text.

Tip No. 73

Tale of a foursome pub-crawl – can peaks elute before the front?

Problem/Question

We all know the meaning of dead time t_M (solvent front, breakthrough time, solvent peak, air peak). It means the time a substance takes to travel from the injection point to the detector – unless it is caught up on the stationary phase. This could be happening to a sample compound that shows inert behaviour within the given chromatographic system. Its elution time should therefore coincide with the elution time of an eluent molecule, and no substance should elute before t_M. No substance could be faster than, say, a methanol molecule – or could it?

Solution/Answer

Yes, peaks eluting before the front are indeed possible, and they are real sample components. We are not talking about ghost peaks, memory effect, peaks from a previous injection or contamination from an eluent, etc.

Explanation:
An inert molecule is a molecule that can diffuse into the pores of the packing without interacting with the stationary phase. It simply diffuses out of the pores again, into the next one and so on and elutes at t_M. If it is a large molecule such as a polymer, it cannot diffuse into the pores. It is excluded and has to remain in the space between the particles known as interstitial volume. As a consequence, it elutes before t_M, in what is called interstitial retention time t_i, see Figure 73-1.

Some of you may now say, "Well, that's all very interesting, but it does not concern me. I am only dealing with small molecules." Nevertheless, you might come across some form of exclusion. Some strongly ionic substances such as salts with their high charge density or bases with their positive charge in a neutral or acidic eluent are downright rejected by silanol groups, due to the Donan potential (like charges repel each other). Thus, even small molecules may elute before t_M. See Figure 73-2, where NaCl elutes before the dead time.

The two bases in Figure 4-1, Tip No. 4, also elute before uracil. If you work frequently with strongly ionic substances, this may happen quite often.

Conclusion

The first early peak in a chromatogram is not necessarily the inert peak – in other words: an elephant, a quarrelsome person, a tee-totaller and a sociable individual go pub-crawling together and make arrangements – just in case – to meet up again in the hotel at the reception desk (our detector). The elephant does not fit through any door of the pubs, and the quarrelsome person cannot get past the bouncers. The poor souls are soon back in the hotel (t_i). The tee-totaller is allowed into the pub, but he does not want to drink, so he comes back at midnight (t_M). The sociable person is having a whale of a time and does not come back to the hotel until 3 a.m. (t_R) – well, perhaps not in the best of shape, but still recognizable.

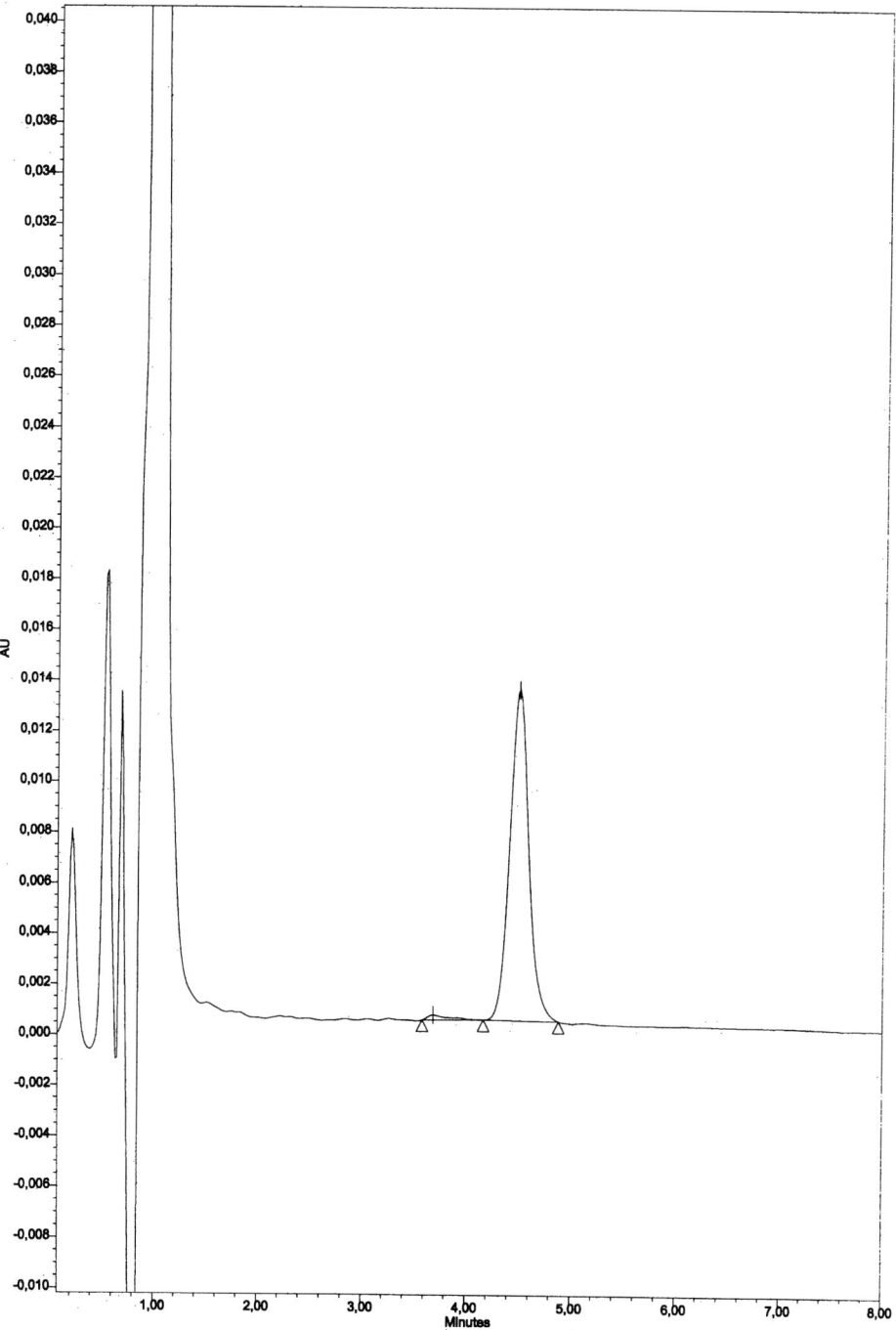

Figure 73-1. Elution of polymer components before the dead time in an RP system. Comments see text.

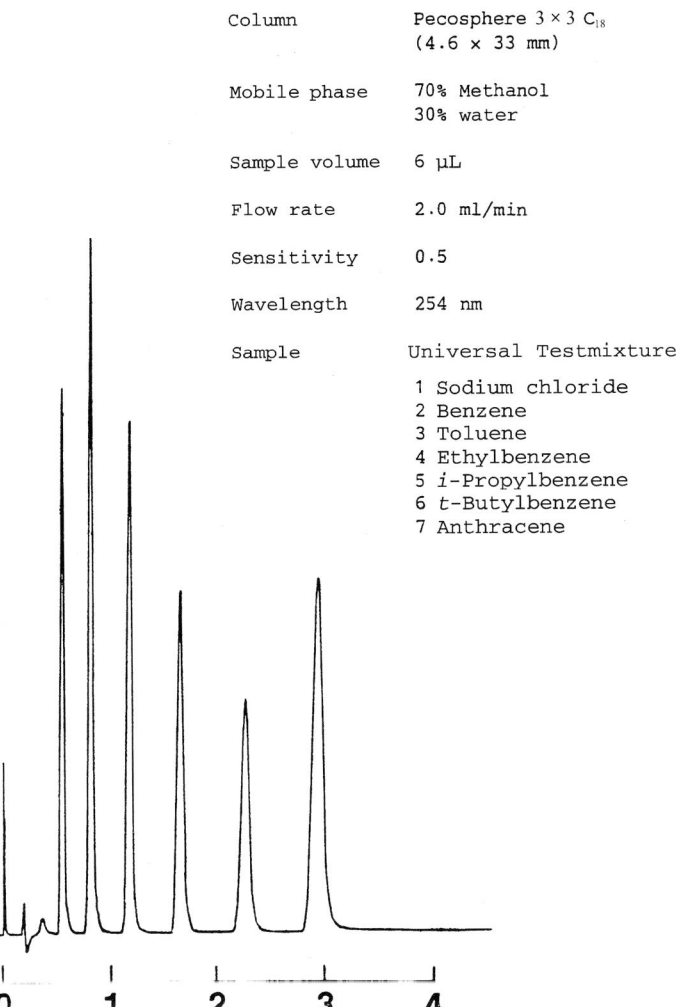

Figure 73-2. Elution of salt before the dead time in an RP system. Comments see text.

References to HPLC-Tips

1. Uwe Neue, *HPLC-Troubleshooting*, American Laboratory and Waters Corporation.
2. Stavros Kromidas (ed.) *Optimization in HPLC*, Wiley-VCH in preparation.

And for those of you who read German:

3. Stavros Kromidas, *Eigenschaften von kommerziellen C_{18}-Säulen im Vergleich*, Pirot Verlag, Saarbrücken, **2002**.
4. Stavros Kromidas, *Handbuch Validierung in der Analytik*, Wiley-VCH Verlag, Weinhein, Reprint **2003**.

2 LC-MS Coupling, Micro- and Nano-LC, Quantification

2.1 LC-MS Coupling

Friedrich Mandel

LC-MS – The one and only universal tool?

If you are already using an LC-MS system or are planning to do so you must have good reasons for it, and buying one does not mean you should discard "traditional" optical detection techniques altogether. While detection limit, selectivity and strength of evidence of LC-MS(/MS) may be unequalled, the initial investments in equipment and time are considerable before a rock-solid LC-MS method can be established. In general, all LC-MS detection methods are not as linear as UV detection and they provide a smaller dynamic range and lower reproducibility. If you determine compounds in complex sample matrices, you should give some thought to interferences of the sample matrix and the analyte signal.

Why this substantial investment in manpower and equipment? It will save you time when developing new methods of detection. This is not a contradiction to what has been said above because while it may take longer to develop a detection method, this extra investment is compensated for many times over by the resulting simplification of sample preparations and chromatographic separations. LC-MS is the method of choice when it comes to quick chromatography of complex samples. Those of us who have spent most of their career perfecting the art of baseline separation will find that LC-MS means a change of paradigm – the quantification of co-eluting peaks. This means that the buffer system has to be chosen and optimized with respect to the detector rather than chromatographic separation. Compared with UV detection, LC-MS offers not only more rapid separations but also better detection sensitivity. Fragmentation patterns provide the possibility for unambiguous identification of substances. However, the LC-MS spectral libraries are still in their early stages – what is lacking is a standardization of instrumental parameters.

Similar to driving a car, where you won't reach your destination without having considered some of the rules of physics, you cannot "just start" with LC-MS. In the following LC-MS tips we will show you how to find the LC-MS technique that best fits your analytical problem. You will also learn about the strengths and weaknesses of the various LC-MS techniques as well as to recognize possible sources of error and how to minimize them. As you may already know – the LC-MS technique that fits everything does not exist, unfortunately. You have the opportunity to choose from various ionization techniques and mass analysers, the application ranges of which mostly overlap. This is why we will compare the most common ionization techniques and discuss their advantages and disadvantages in the following sections. No need to be scared, this will not end up in a lesson on ion physics – it will be a rather pragmatic approach to questions such as, "How does the charge get on my molecule?" and "How can I detect the charged molecule?" My intention is to describe not only

Figure 2-1. Photograph of John Fenn.

"how it works" but also "why", in order to enable you to develop and optimize your methods on your own, without having to consult books or ask for assistance most of the time.

LC-MS is no longer the exotic technology it once was, and some of the mass spectrometers available nowadays hardly take up much more bench space than an HPLC system. Although it first appeared in the seventies, it was only in the early nineties that instruments and ionization techniques suitable for routine work became widely available. The big breakthrough happened with the use of atmospheric pressure ionization techniques, which earned John Fenn the Nobel Prize for Chemistry in 2002. LC-MS can be used to quantify trace amounts, collect fractions based on mass, elucidate molecular structures by MS^n, identify very small amounts of proteins by using nano-HPLC techniques, and much more. Today the user interfaces of the data systems are very easy to operate, including "open access" or "walk-up" mass spectrometry, and for many users, the LC-MS system seems to have become a "black box". In the following tips, I invite you to have a look into the inner workings of your LC-MS system.

Tip No. 74 — Choosing the right LC-MS interface

In all decisions you take in LC-MS, the most important criterion is ionization. Whatever you may undertake in order to achieve highly sensitive, highly selective and/or high-resolution measurements – it is most important to ionize the analytes as gently as possible. A mass spectrometer is only capable of measuring either positively or negatively charged molecules ("pseudomolecular ions"). Unfortunately, very often, the chemistry of these ions doesn't make it easy for you – the analytes to be measured can cover a wide range of polarities and molecular weights. They may also be more or less thermolabile (otherwise we would use GC-MS instead of LC-MS). In addition, a mass spectrometer has to be operated at high vacuum. Therefore, we first have to evaporate the mobile phase and then remove it from the analytes ("desolvation"). As we will see in the following sections, some LC-MS interfaces are capable of handling high LC flow rates, while others show their strengths at low flow rates. Whatever technique you choose, it will be an API technique (atmospheric pressure ionization), such as API-electrospray (ESI), atmospheric pressure chemical ionization (APCI) or atmospheric pressure photoionization (APPI). All other interfacing techniques (thermospray, particle beam) are either just of historical interest or they are not suitable for routine work.

**Polar, thermolabile analyte, high molecular weight –
shall I use capillary or nano-HPLC?**

Try electrospray first! Before you begin you should understand how it works. In fact, electrospray is not a method for ionizing analytes, but a way of releasing cations

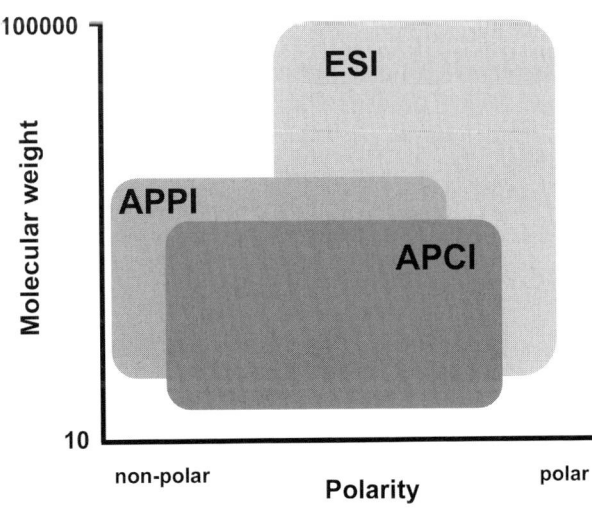

Figure 74-1. Application ranges of API techniques.

or anions that have been formed in the HPLC buffer system by the acid/base equilibrium. This aspect is of fundamental importance when you develop your LC-MS methods or when you are troubleshooting with your system. You should therefore choose pH conditions that force the formation of free analyte cations or anions in solution. The selection of "MS compatible" mobile phases will be the topic of one of the tips to follow.

API-electrospray usually utilizes a pneumatic nebulizer in order to convert the eluent into an aerosol. This is also called "pneumatically assisted electrospray" or "Ion-Spray®". The aerosol is then sprayed into an electrostatic field. A high voltage of several kilovolts is applied between the nebulizer and the entrance of the ions into the vacuum system ("orifice"). Depending on the polarity of the charged analyte molecules, the resulting aerosol droplets carry a positive or negative net charge. This causes them to be attracted towards the orifice or the ion transfer capillary where they are transferred to the ion optics of the mass spectrometer. It is important not to introduce the charged droplets into the mass spectrometer. This would trigger a high background signal and therefore negatively affect detection sensitivity. Modern ion source designs are therefore based on an orthogonal geometry, spraying the eluent at a 90° angle onto the ion entrance and the ion path. Simultaneously, heat is applied via a stream of heated nitrogen, in order to evaporate the solvent from the aerosol droplets. On their way towards the orifice or ion transfer capillary, the droplets undergo a cascade of shrinkings and explosions into smaller droplets ("Coulomb explosions") in order to stabilize the droplet's surface charge. Also (pseudo-) molecular ions are spontaneously emitted from the droplets. Orthogonally designed ion sources ensure that almost exclusively desolvated (pseudo-) molecular ions find their way into the mass spectrometer. Because of their high mass, residual droplets cannot be deflected by

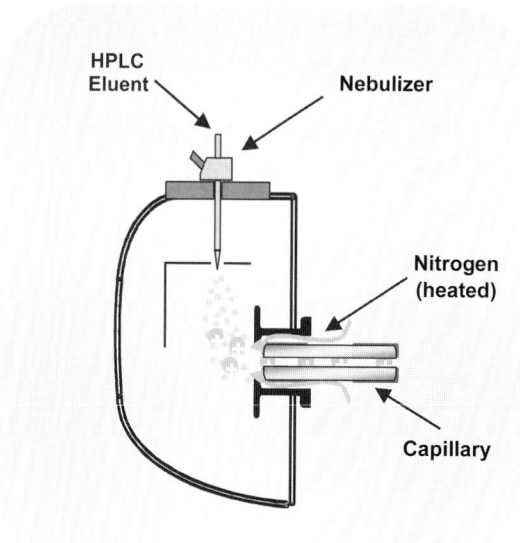

Figure 74-2. Schematic representation of an API-electrospray ion source.

90°. Orthogonal ion sources not only exhibit a high level of sensitivity, but also of robustness and are only marginally affected by contamination.

Which flow rates can be handled by electrospray ion sources? With pneumatically assistance operation is in a range of 0.5 µL min^{-1} to 2 mL min^{-1}. Of course this could involve a variety of geometries and/or nebulizers. For flow rates of 0.5 µL min^{-1} to ca. 50 µL min^{-1} nebulizers are used that have been optimized for extremely low dead volumes ("microspray"), while a standard nebulizer is used for higher flow rates. Depending on the manufacturer, these provide optimum ion yields within a flow rate range of 100 µL min^{-1} to 1 mL min^{-1}. We already know that in electrospray the ions are formed in solution. Therefore, it is a concentration-dependent detection technique, as in UV detection. This has the advantage that the eluent stream can be split in front of the mass spectrometer without losing sensitivity. This is very useful when using HPLC columns with large diameters and the high HPLC flow rates associated with them, when using two detectors simultaneously or when MS-based fraction collection of the eluent stream is required ("mass-directed" or "mass-based" fractionation).

Very low sample amounts and volumes, as in "proteomics" applications, require a highly sensitive method of detection. Currently, the most sensitive LC-MS technique uses a nanospray in conjunction with nano-HPLC. Nanospray is actually a variation of the original "classical" electrospray, where the aerosol is formed solely by high electrical fields, without any pneumatic support. At the moment the use of nanospray is restricted to the detection of biopolymers (peptides, proteins, oligonucleotides). With increasing robustness it could also be transferred to other application areas.

An important characteristic of API-electrospray is formation of multiply charged (pseudo-) molecular ions – depending on the number of basic (positive ESI) or acidic (negative ESI) functional groups and the HPLC buffer system. Why is this so important? Any mass spectrometer has a limited (physical) mass range. The upper limit is usually around m/z 4000 to m/z 6000. The exceptions are time-of-flight mass spectro-

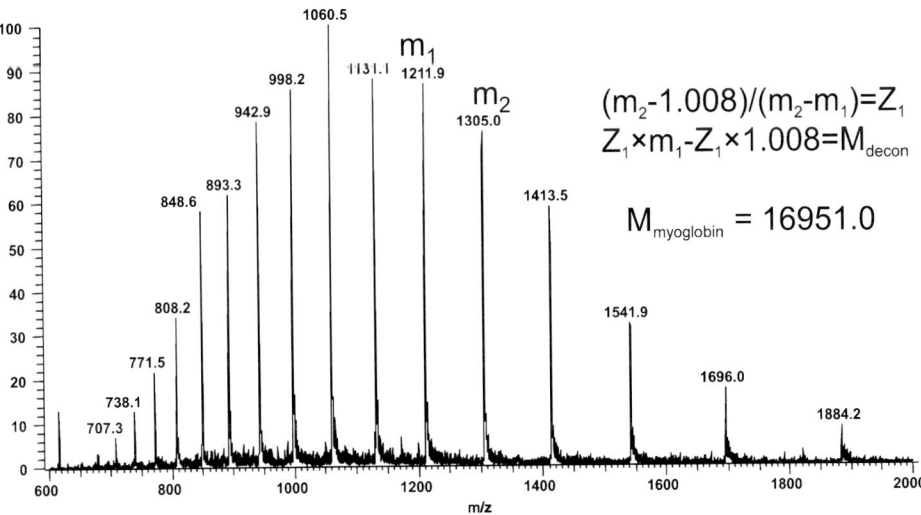

Figure 74-3. Electrospray mass spectrum of myoglobin.

meters with high acceleration voltages. You probably know that in mass spectrometry it is always the mass/charge ratio that is determined. Only the ability of electrospray to form multiply charged ions enables us to measure proteins of molecular weights above 100 000 Da. Figure 74-3 shows the mass spectrum of a small protein. Each peak represents the various charged states of the molecular ion.

Low polarity, high HPLC flow rates, volatile analyte?

In this case, APCI is your interface of choice. The difference in construction between APCI and API-electrospray is minimal, however essential for the ionization process. Unlike in electrospray, we do not make use of the ions formed in solution, but generate them in the gas phase instead. The aerosol formed from the mobile phase and the analyte is sprayed into a heated ceramic or glass cartridge and completely evaporated. That vapour is sent past a needle, to which several kilovolts are applied, and at the tip of which an corona discharge is formed. The plasma produced in the process enables chemical ionization to take place. While colliding with the (still) neutral analyte molecules, the ionized solvent molecules act as proton donors or acceptors. The difference in proton affinities between the molecules colliding with each other is important. With electronegative analyte molecules we can also observe electron capture processes. As the charges are transferred to the analyte molecule through single collisions, we only generate singly charged molecular ions in APCI. Of course, this limits the molecular weight range that is detectable compared with electrospray. A further restriction in the application of APCI is the need to evaporate the analyte without thermal degradation – a step many thermolabile compounds do not survive. Therefore, the vaporizer temperature is an important parameter in APCI method development.

What are the reasons in favour of APCI as a detection technique in LC-MS? It is ideally suited for high HPLC flow rates. Most APCI source designs reach their optimum performance at 0.5–1.5 mL min^{-1}. A great advantage of APCI is its ability to

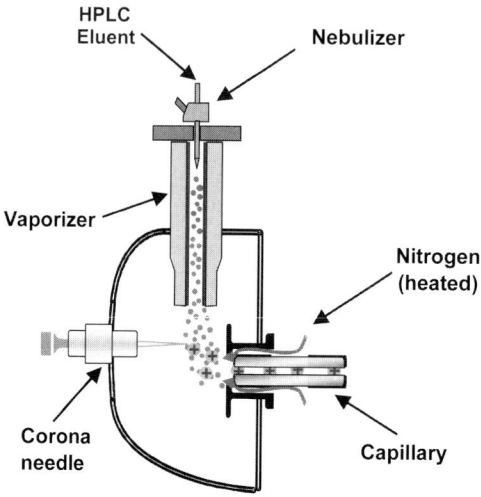

Figure 74-4. Schematics of an APCI ion source.

ionize analytes of weak polarity, which are not accessible in API-electrospray at all or only under extreme pH conditions. In APCI the decisive factor is not pH value of the mobile phase, but the gas phase acidity/basicity. This allows you to optimize the pH of the mobile phase solely for chromatographic separation without also having to consider the mass spectrometer. In addition, in APCI you will not find "mixed ionization" caused by protonation and alkali adduct formation. As we will see in one of the subsequent tips, APCI methods show a more linear response than API-electrospray methods and also suffer less from ion suppression effects.

Still no signal, analyte difficult to evaporate or non-polar, APCI not sensitive enough?

Although it is not yet a commonly used method, APPI could be the answer to your problems in analysis. Atmospheric pressure photoionization is a slight modification of APCI. After the evaporation step, the analyte and the mobile phase are sprayed past a krypton UV lamp emitting light energy of 10 and 10.6 eV. This excitation energy induces photoionization of the analytes, the ionization energy of which, of course, has to be below 10 or 10.6 eV, respectively. The widely used HPLC mobile phases, such as water, methanol, acetonitrile and hexane, exhibit ionization energies above the excitation energy and will thus not be ionized. This first photoionization step leads to the formation of a radical cation, to which a hydrogen atom originating from a non-ionized solvent molecule may be transferred. Sometimes, the radical cation and the molecular cation can be detected alongside each other. Should the detection sensitivity of this direct photoionization not be sufficient, you can easily add an ionizable modifier ("dopants") to the mobile phase directly in front of the ion source (about 5% v/v to the eluent). The photoionized dopant then serves as a proton donor, ionizing the analyte. Typical dopants or modifiers are acetone and toluene. Acetone is also an excellent donor of thermalized electrons, which can be used for analyte anions after

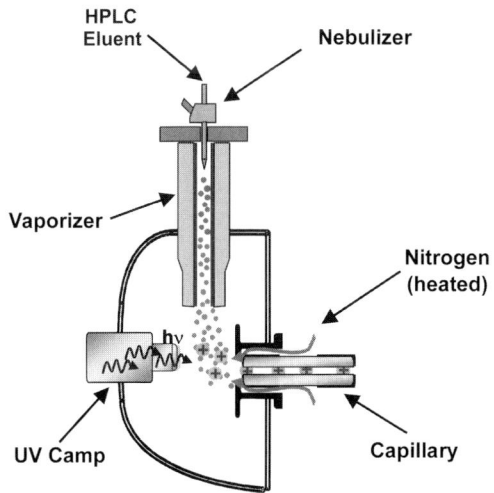

Figure 74-5. Schematics of an APPI ion source.

an electron capture step. The optimum HPLC flow rates for APPI are in the range of 0.5 mL min^{-1}.

It is amazing which non-polar compounds can be analysed using APPI. It is the one and only ionization technique that enables the sensitive formation of polycyclic aromatic hydrocarbon cations. In contrast to APCI, the ionization process in APPI is not negatively affected by residual aerosol droplets. Because of this, the vaporizer can be operated at lower temperatures. Of course, this helps in the detection of thermolabile compounds that are not detectable by API-electrospray. In general, the sensitivity of APPI is fairly comparable to APCI. However, in APPI the range of analyte polarity and thermolability is significantly wider than in APCI.

Tip No. 75 — Which mobile phases are compatible with LC-MS?

You may have anticipated it already – that as an LC-MS user you will have to say goodbye to a good old friend – the phosphate buffer. As you will see in a minute, there are exceptions to any rule. However, when converting existing HPLC methods into LC-MS or by developing new methods, you should base all of them on volatile buffer systems. Before we start discussing buffers – let us talk about which solvents work best with which LC-MS interface.

The solvents

The following solvents are compatible with API-electrospray and APCI: alcohols, acetonitrile, tetrahydrofuran (THF), water, acetone, dimethyl formamide, methylene chloride and chloroform. If, in APCI, you at least partially replace acetonitrile with methanol, you will enhance detection sensitivity as well as the long-term stability of the analyte signal. This is because gaseous acetonitrile is a relatively strong base and therefore competes with the analytes for protonation. Additionally, acetonitrile tends to polymerize in the APCI plasma, coating the corona needle with an insulating layer after some hours of operation. Consequently, more frequent abrasive cleaning of the APCI needle will be required. Dimethyl formamide should be below 10% v/v when running API electrospray, while in APCI you should be prepared for a high background signal. In APCI, THF also tends to polymerize, in particular when it contains traces of peroxides. There is a trick to minimising the precipitate in the APCI ion source. Just before the mobile phase enters the ion source, you should add about 5% v/v of water. This also stabilizes the corona discharge while running THF. Halogenated hydrocarbons can enhance the ion yield when used as modifiers in APCI. In API-electrospray they show neither a positive nor a negative influence. In general, the less protic your solvent is, the less it will be suited for API-electrospray (acid/base equilibrium).

In APCI we make use of the mobile phase as a "reagent gas" for chemical ionization of the analyte. Therefore, besides the ones already mentioned, aliphatic and aromatic hydrocarbons are also permitted, as well as CS_2 and CCl_4. Toluene is an excellent proton donor in APCI.

The additives

What should be used instead of phosphate buffers? The most important rule is to use volatile buffer additives and to use organic acids. I can hear the screams of protest of the experienced HPLC user. But, to be honest, most of the analytes can be separated sucessfully in an "MS-compatible" way by using the modern RP column materials. Please take into account that mass spectrometers are able to differentiate between compounds by the *m/z* signal if they cannot be separated chromatographically. For your validated methods that use phosphate buffers, I will tell you what the exception from that rule is.

However, let us discuss the additives first:

Use ammonium acetate or formate to buffer the pH. Acidify the pH by adding acetic acid, formic acid, trifluoroacetic acid (TFA) when running positive ion electrospray and adjust to a basic pH in negative ion electrospray by adding ammonia, triethylamine (TEA) or *N*-methyl morpholine. Another rule is to use a buffer concentration as low as possible – below 10 mmol L^{-1} in electrospray and maximum 100 mmol L^{-1} in APCI or APPI. Please take into account that TFA is a weak ion-pairing reagent and therefore reduces the detection sensitivity for many analytes. So avoid the use of even low TFA concentrations if you frequently change ion polarity in electrospray. The TFA anion gives a permanent background signal at *m/z* 113. If you need a basic pH for either the chromatographic separation and/or the ionization, you should use ammonia instead of TEA. Ammonia will not show any memory effect in your HPLC system while TEA shows a background signal at *m/z* 102 during subsequent measurements in positive ion electrospray. Both with TFA and TEA it could take days until the background signal has dropped to an acceptable intensity level.

Why all that worry about volatile buffers and low buffer concentrations? After evaporation of the solvents, non-volatile buffers will precipitate inside your ion source and, depending on the ion source geometry, will soon block the ion entrance or will cause current leakage or shortings. Even if this does not occur, the alkali cations will block the aerosol droplet surfaces (potassium or sodium phosphate) in electrospray ionization and hinder the emission of the analyte ions. The need for low buffer concentrations is also easily explained. With all API techniques, a cloud of charged species is formed together with the ionized sample matrix and buffer additives. The charge density in this cloud is limited by space charging (distraction) of species of the same polarity. Finally, this leads to a dispersed spray and a "dilution" of the analyte ions in the cloud at higher buffer concentrations. As the spray is more collimated in API-electrospray than in APCI or APPI, the latter two are more suited to high buffer concentrations.

Tip No. 76 — Phosphate buffers – the exception

Please do not interpret the following hint as a general licence to follow old habits of working with phosphate buffers. Do not deviate from the iron rule of developing any new LC-MS methods by using volatile buffers. But what if you have to investigate a chromatographic peak – and unfortunately the method had been extensively developed and/or been validated using phosphate buffers? In this case, you might be allowed to work with the "forbidden" buffers. However, you should be prepared for a heavy contamination of the spray chamber and a major cleaning-up effort. Many ion source designs will allow you to keep your tools on hand. Please also consult the manufacturer of your mass spectrometer or other users of the same instrument type.

How does it work?

Remember the first tips. If you want to determine analyte cations, then electrospray suffers from a dramatic loss in detection sensitivity with sodium or potassium phosphate buffers. The reason for this is the suppression effect caused by the alkali cations in the aerosol. Therefore you should select APCI as an ionization technique while using these non-volatile buffers. In APCI, both analyte and mobile phase are evaporated before the ionization step. Of course, non-volatile buffer components will precipitate as a white powder in your ion source. Nevertheless, the analytes will be protonated in the corona discharge. Because of the severe contamination of the ion source, your LC-MS system will "survive" this procedure for only a few hours or even just for a single chromatographic run, depending on the buffer concentration. When looking for analyte anions, however, you should choose API-electrospray instead of APCI. The reason is that the phosphate anion is volatile enough to be eliminated from the aerosol droplets, similarly to the analyte anions. However, please be aware of some reduction in sensitivity. The mass spectrometer will "forget" the phosphate buffer treatment rapidly. Your HPLC system and your column however will deliver alkali cations even weeks later, causing unwanted and often very disturbing adduct formation.

Summary

In few exceptional cases (!) you can operate your LC-MS system with non-volatile buffers. Use APCI for the detection of cations and API-electrospray for the detection of anions. Be prepared for the rapidly increasing contamination of your ion source as well as for the frequently required cleaning procedures. Preferably use a separate HPLC system with non-volatile additives.

Tip No. 77 Paired ions

Besides phosphate buffers, there are other troublemakers which can have a comparable impact – the ion-pair reagents. As with non-volatile buffer systems, ion-pair reagents significantly reduce the sensitivity of detection. Don't get confused by publications stating the opposite – if analytes do not form ion pairs, naturally their ionization will not be negatively affected. But why do you want to use this type of buffer additive? You might want to make highly polar compounds more lipophilic rather than ionic in the mobile phase. The effect is an increased retention on reversed-phase columns and less peak tailing. However, as you know already, electrospray ionization is based on the release of ions from the aerosol droplets. When masking the analyte ion by a counter ion you will not be able to detect your analyte, unless you break the ion pair apart through the application of heat.

Which "antidote" is available?

You should avoid strong and non-volatile ion-pair reagents, such as tetrabutylammonium bromide or heptanesulfonic acid. Not only does their usage drastically reduce sensitivity, being themselves ions in solutions, but they contaminate every HPLC system long term and cause a high background signal. For example, even ultra-low traces of the tetrabutylammonium cation result in an intense signal at m/z 242. The weak ion-pair reagent TFA, which is widely used in peptide analysis, will significantly suppress the signal of basic compounds, such as the LSD metabolite LAMPA. As shown in Figure 77-1, LSD itself is not negatively influenced by TFA.

You can counteract the suppression effect of TFA by adding a high concentration of organic acid (i.e., 50% v/v of propionic acid in isopropanol) immediately before

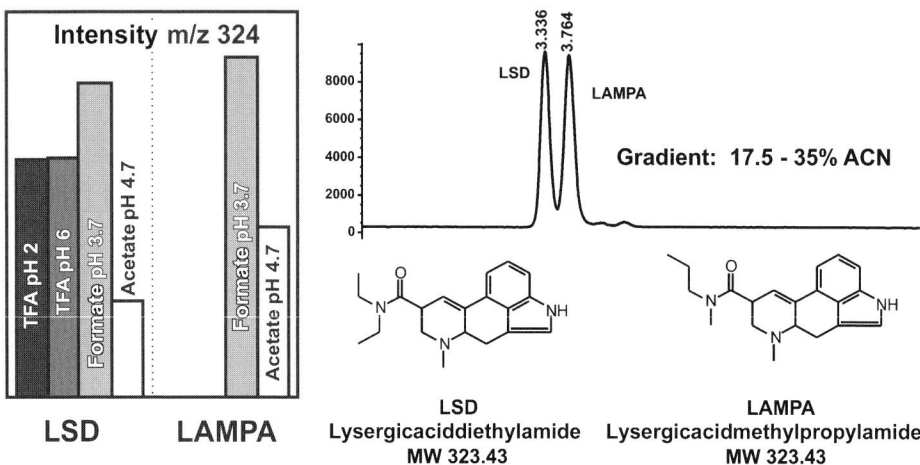

Figure 77-1. Influence of TFA on LSD and LAMPA.

the electrospray ion source, in order to displace the TFA anion from the ion pair. If you cannot avoid the use of ion-pair reagents, you should take their volatility into account. Acidic functional groups can be appropriately masked by aliphatic amines, such as triethylammonium acetate, n-butyl-dimethylammonium acetate or di-n-butyl-ammonium acetate. On the other hand, you can replace alkanesulfonic acids with perfluorinated organic acids of different aliphatic chain lengths. In most cases, even in API-electrospray, there will only be very little suppression of detection sensitivity. If possible, you should opt for APCI or APPI, as the weak ion pairs just described will be cleaved by the evaporation process. Please be prepared for a significant background signal when you invert the ion polarity after your ion-pairing experiments. Just as when using high alkali concentrations – i.e., phosphate buffers – you should mark the bottles and HPLC columns used with the ion-pair reagents and never ever use them with "normal" mobile phases. Moreover, it is not a good idea to put these "contaminated" mobile phases through a vacuum degasser, which typically has a large inner surface consisting of porous Teflon and would thus produce an unpleasant memory effect.

Summary

Traditional ion-pair reagents are not suited for use in LC-MS. It is better to use volatile reagents such as aliphatic amines or perfluorinated organic acids at the lowest possible concentration. Avoid contamination of your vacuum degasser and use solvent bottles and HPLC columns that are dedicated to use with ion-pair chromatography.

Tip No. 78 Using additives to enhance API-electrospray ionization

In LC-MS, as in everything else in life, it is the dosage that matters. You can enhance ionization and detection sensitivity by adding Na or K ions to the eluent, as long as you obtain the correct concentration inside the ESI ion source – it should be in the region of 0.5 mmol L^{-1}. Please make use of post-column addition and do not add the modifier to your mobile phase upfront. This will avoid unwanted surprises by contamination of your HPLC system.

Formation of cations through the addition of alkali salts is ideally suited for all compounds that contain several OH functional groups, such as carbohydrates or steroids. These are difficult to protonate and thus not very sensitive to electrospray detection. By adding modifiers you obtain a uniform ion formation and therefore a higher sensitivity and increased reproducibility. Figure 78-1(a) shows that prednisolone produces multiple molecular cations, even when acidified with TFA. The intensity is distributed across many molecular cations, the ratio of which can vary significantly depending on the mobile phase composition. Therefore reproducible quantitation is almost impossible. In Figure 78-1(b) an equimolar mixture of different modifiers was added in order to evaluate their affinities to prednisolone. It can easily be seen that the Na adduct provides the most intense signal. Figure 78-1(c) shows the result of post-column addition of sodium acetate. Now we have a uniform ionization and a stable pseudomolecular ion. But it is not only alkali salts that make carbohydrate detection easier in ESI. You can also add 50 mM HCl post-column (concentration in the ESI source about 2 mmol L^{-1}) and force the formation of chloride adducts. These are of course detected in the negative-ESI mode.

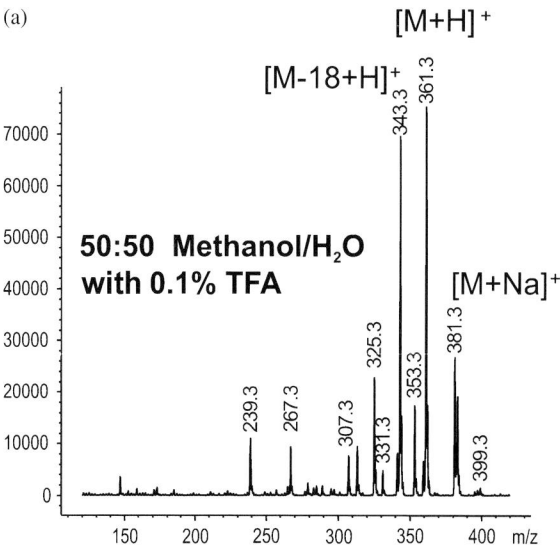

Figure 78-1. Adduct formation in ESI.

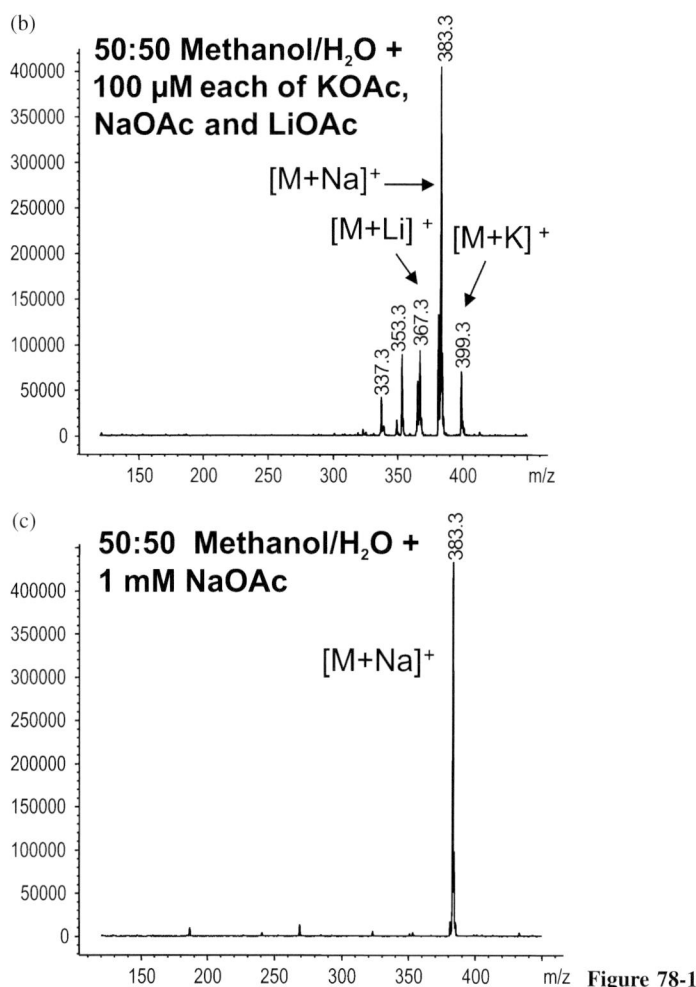

Figure 78-1.

Additives for APCI

Let us stick to the prednisolone example. Normally it gives an $[M + H-H_2O]^+$ signal in positive ion APCI, while not being detected in negative ion APCI. However, if you add 1–5% v/v methylene chloride post-column, you obtain a very intense $[M + Cl]^-$ ion (Figure 78-2). Phenolic compounds can be detected with enhanced sensitivity in negative ion APCI as soon as traces of either oxygen or trichloromethane are present. A very common APCI modifier is toluene, which, as a post-column addition to the eluent (5 % v/v), makes an excellent proton donor.

Summary

The possible additives range from alkali salts and methylene chloride to oxygen. It is up to your own creativity to find your personal "secret recipe". Simply try additives that you know to have high affinities for your analyte's functional groups or that form reactive species in the APCI plasma.

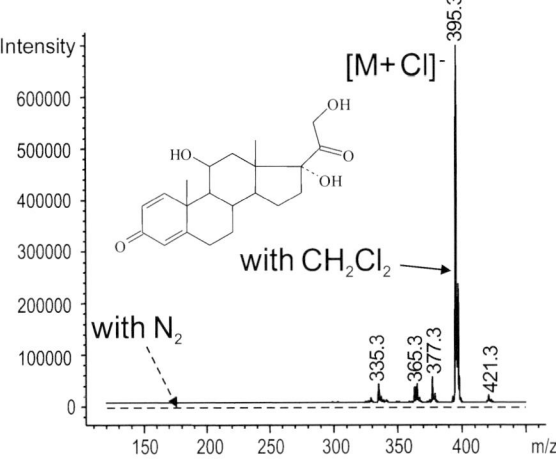

Figure 78-2. APCI adducts of prednisolone.

Tip No. 79 How can I enhance sensitivity of detection?

Think through all your method parameters step by step. Did you choose the right ionization technique? Usually ESI is more sensitive than APCI or APPI. However, if for chromatographic reasons you must select a mobile phase where your analytes are not ionized in solution – you must resort to one of the chemical ionization techniques. But let us assume that you selected an ionization technique that is suitable for your analyte.

Electrospray

The optimal pH value is two units below (positive ion ESI) or above (negative ion ESI) the pK_a value of the compounds to be detected. In this case, more than 99% of your analyte molecules exist as ions in solution. Maybe you have to add an acid or base post-column in order to achieve these conditions. Some ESI nebulizers generate aerosol droplets that are too large when spraying 100% aqueous mobile phase. You should then apply more heat, for example, by applying "heated drying gas" or "turbo ion spray", or increase the percentage of organic solvent in the eluent by adding methanol or isopropanol post-column. If necessary, combine this with adjustment of the pH value or the addition of modifiers. Is your background signal significantly stronger than your analyte signal? Then presumably your ESI detection will suffer from ion suppression. Eliminate the reasons for the high background signal. Try another batch or brand of your HPLC solvents. "HPLC grade" means that this particular solvent is suitable for UV detection, but it does not necessarily mean it is LC-MS compatible. *m/z* 102 in positive ion ESI results from previous measurements using triethylamine, *m/z* 279 is most probably caused by dibutyl phthalate, while a series of peaks at intervals of *m/z* 44 result from ethoxylated surfactants. If you still use an HPLC column with a large internal diameter, you had better change to smaller column dimensions (3 or 2.1 mm ID). Please remember – ESI is a concentration-dependent detection technique. Therefore, the detection sensitivity increases quadratically with a reduction in the column diameter. In addition, the associated lower HPLC flow rate also contributes to the sensitivity enhancement.

APCI

In APCI you should check the vaporizer temperature. Two extremes could cause loss in sensitivity – temperatures that are too elevated induce pyrolysis, while too low temperatures result in incomplete evaporation of the analyte. Do you generate sufficient APCI reagent plasma? Try to increase the corona current in small steps. In most cases this increases not only signal height but also signal stability. Is your HPLC flow rate high enough? As APCI is a mass flow-dependent detector you gain sensitivity by increasing the column diameter, which increases the HPLC flow rate. Do not set the flow rate too high though – the optimum for most available APCI ion sources lies in the range from 0.8 to 1.5 mL min^{-1}. If your compounds exhibit lots of OH functional groups you could also try the additives that were discussed in the previous chapter.

APPI

The optimal flow rate for APPI is around 0.6 mL min^{-1}. If your analyte undergoes direct photoionization (APPI without dopant) then APPI is fairly concentration dependent (Lambert-Beer's law), if you use a dopant (acetone or toluene) then APPI behaves like a mass flow-dependent detector. You should adapt your chromatographic conditions accordingly. In addition, the vaporizer temperature and the capillary or cone voltage influences APPI sensitivity. Unfortunately it is virtually impossible to predict the analyte behaviour – you have to determine optimum APPI conditions empirically.

Optimizing instrument parameters in APPI

As the lamp intensity cannot be changed by the operator, the vaporizer temperature, the temperature of the added nitrogen stream and the capillary or cone voltage are options for optimization. If you cannot generate sufficient ionization by direct photoionization you should try dopant APPI. For this purpose, you add 1–5% v/v toluene or acetone to the eluent (T-piece immediately in front of the ion source). Acetone is the best electron donor available for negative ion APPI. Please check on the sensitivity by running real samples, not standard samples. In APPI, the absolute intensities often seem to be low. The most important decision criterion, however, is the selectivity of analyte ionization in relation to the chemical background.

Tip No. 80 — No linear response and poor dynamic range?

When switching from optical detection techniques to mass spectrometry you will primarily be concerned with a reduced dynamic range and very often with a non-linear response. I admit that the dynamic range maybe one to two orders of magnitude lower compared with the use of a diode array detector, it is mainly in the range of from 10^3 to 10^4. However, this limitation typically occurs at high concentrations. The solution to this problem is simple – dilute your sample. In LC-MS this is not normally a problem, as this detection technique beats any optical detector in sensitivity. When discussing linearity you should keep in mind that at low concentrations, in most cases, we observe a linear detector response. Again, your first option would be dilution of the sample.

The reasons

As we have learned already, we generate a high charge density in the spray, in particular when using electrospray. At injected concentrations of about 0.5 mg mL^{-1}, the ion source is affected by what is known as the space charge limit effect, i.e., the charged cloud before the orifice diverges with higher concentration, but the number of ions that reach the mass spectrometer will not increase. What is particularly annoying is that it is mainly the compounds with a high ionization yield that are affected by this phenomenon, also known as "soft clipping". A classical example is shown in Figure 80-1. While *o*-toluidine exhibits a perfectly linear response in ESI, 3,3-dimethyl benzidine – being simply double the *o*-toluidine molecule – shows a quadratic behaviour. The duplication of the ionizable functional groups per molecule increases the ionization yield and therefore also limits the dynamic range.

Possible solutions

Use isotope-labelled internal standards – if possible at a concentration below the "critical" analyte concentration. The space charge effects now influence both the analyte and internal standard. It would be even better if you switched from ESI to APCI or APPI detection. In both techniques the cloud of charged species is less dense. Because of this, the non-linearity problems are significantly shifted to higher concentrations. The response curves in APCI and APPI are generally more linear than in ESI. For example, our case of 3,3'-dimethyl benzidine shows very good linearity in APCI (see Figure 80-2).

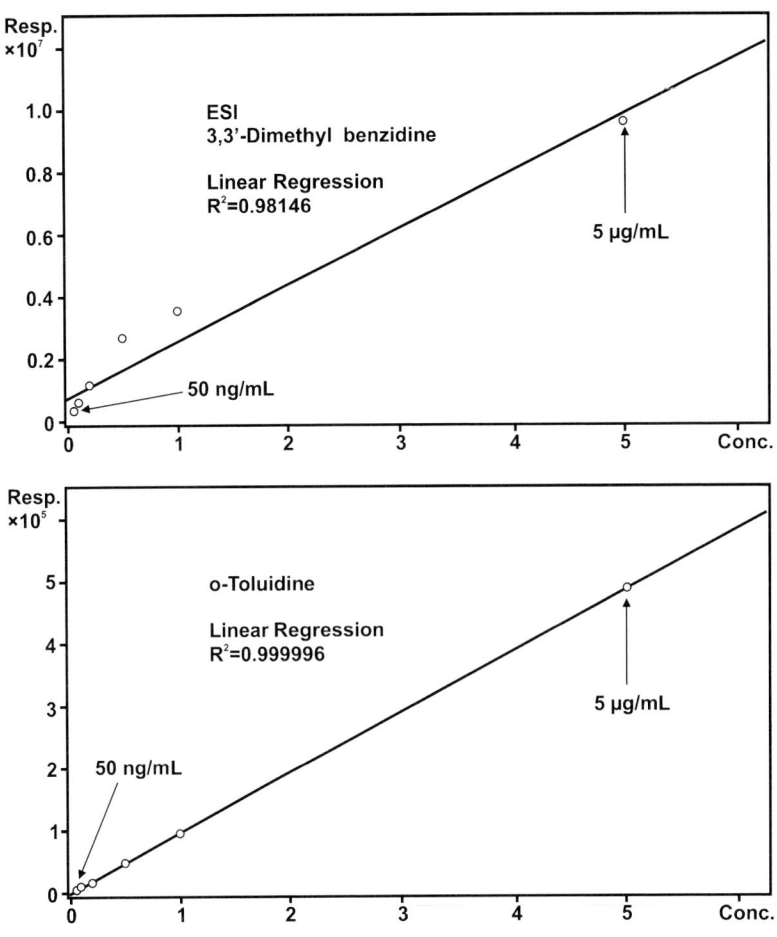

Figure 80-1. ESI calibration curves of *o*-toluidine and 3,3′-dimethylbenzidine.

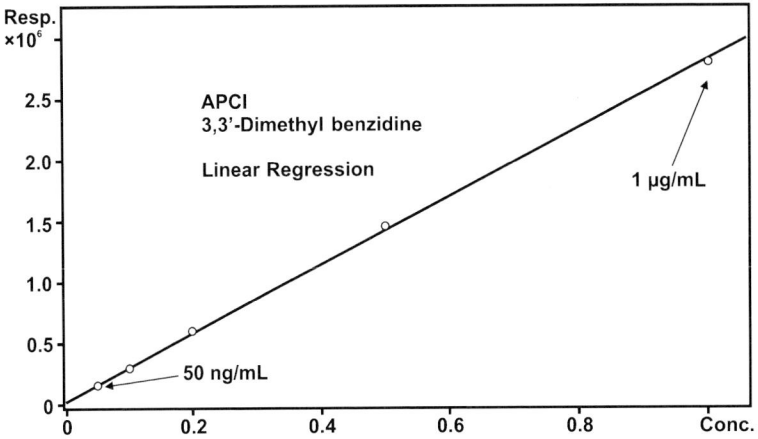

Figure 80-2. APCI calibration curves of *o*-toluidine and 3,3′-dimethylbenzidine.

Summary

High sensitivity of detection can be a curse. High ionization yields result in space charge effects in the spray chamber and harm both the dynamic and linear range of detection, in particular in the electrospray mode. Very often you can improve performance by simply adapting the concentration range of your samples (dilution). If you have access to an APCI or APPI ion source you should try to change your method from ESI to these more linear detection techniques. Slight losses in sensitivity compared with ESI will be compensated for by a significantly better linearity.

Tip No. 81 — How much MS^n do I need?

There are many reasons to opt for MS^2 or MS^3. Let us focus on decision criteria that are based on analytical reasons. The more complex your injected sample is and the shorter your chromatographic separations, the more specific your detector has to be. This applies to a mass spectrometer the same way as it does to other detection methods. You do not simply selectively measure just the molecular ion or one or multiple fragment ions – the high specificity of MS-MS results from the detection of a transition or a reaction from a "precursor" to a fragment. You detect the origin of the signal, not just its existence. This is why in MS-MS you talk about "single reaction monitoring" (SRM) or "multiple reaction monitoring" (MRM). If you monitor full MS-MS spectra instead of MRM transitions, you lose about a factor of 10 in sensitivity. No sensitivity loss happens by using ion trap mass spectrometers.

Solutions

After having read the previous tips you already know that artifacts (ion suppression, alkali adducts) are caused by the ionization process, regardless of what type of mass spectrometer you use behind the ion source. The need for a reasonable chromatographic separation of the analytes from the sample matrix increases with the sample complexity. If the analysis time is an important factor in your analytical environment, you can hardly work without using an MS-MS capable mass spectrometer. Even if you use quantitative MS-MS methods that have been optimized in each and every detail, the use of internal standards is highly recommended. These internal standards should undergo the same analytical processes as the analytes (stable isotope labelling).

If you are just interested in the mass spectrum or the molecular weight in order to confirm a synthesis step, you could actually work without any chromatographic sepa-

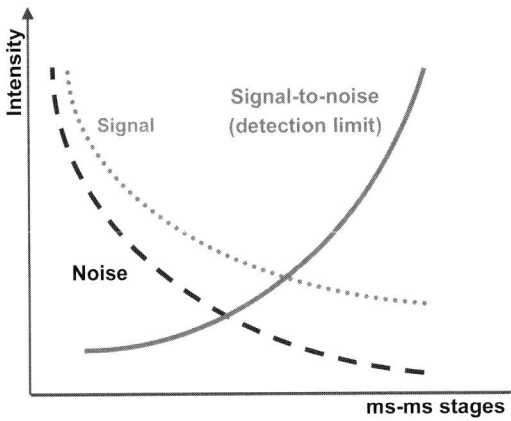

Figure 81-1. Signal-to-noise as a function of MS^n stages.

ration, in a flow injection mode (FIA), and you would not need an MS-MS capable mass spectrometer either. Fragment ion spectra can be obtained by collision-induced dissociation (CID) inside the ion optics – even when using fairly simple mass spectrometers (i.e., single quadrupole or time-of-flight instruments). If you follow a very selective sample preparation procedure and/or if you have developed a good chromatographic separation, you can work with high detection sensitivity and selectivity, without even using MS-MS.

Summary

Do not ask yourself "How much MS^n do I need?" but "How much effort is needed in the chromatographic separation in order to minimize MS-related artefacts?"

Need more help?

What else can you do in order to use LC-MS successfully? Read the manuals that come with your LC-MS system. Here you will find hints on how to enter method parameters, the optimization and tuning. In general the instrument manufacturers provide default methods with their instruments, which appropriate as starting points for your own method development.

It is false economy not to visit the operator training courses your instrument manufacturer offers. There the features of your LC-MS system will be explained to you. "Classroom training" will give you the opportunity to meet other users with similar interests. After the training course, you should keep in touch with the other participants. If you want to brush up on theory – there are manufacturer-independent training programmes where you could become familiar with the fundamentals of LC-MS.

Rather than re-inventing the wheel you should ask colleagues or your manufacturer's application chemists for advice. Application chemists are confronted with a wide range of analytical problems on a daily basis. In most cases they are able to help you with your questions. For troubleshooting, they are also the right people to contact.

Read LC-MS-relevant literature. Using Internet-based search engines you can easily find suitable information. On the websites of the LC-MS manufacturers you can find a variety of examples of applications. Methods that have been reported using your type of instrument are easy to transfer to your set-up. However, please consider the different instrument designs of the different manufacturers – ESI sprayers, for example, can have orthogonal or off-axis geometry, heating could be realized via heated nitrogen or a heated ion transfer capillary, and much more. Increasing a "cone voltage" may have the same effect as increasing a "fragmentor voltage" – more and more fragmentation of the molecular ions. However, the absolute voltage settings may differ significantly. This is just one example of how different approaches (instrument parameters) lead to the same result (mass spectrum).

The "take home message" is that LC-MS(-MS) is simple in principle. Trouble arises because of the ionization chemistry and the fragmentation behaviour of the analytes. Do not give up all too soon – even the so-called "experts" with five years' experience had to go through those five years first.

References to LC-MS

Electrospray Ionization for Mass Spectrometry of Large Biomolecules Fenn J.B., Mann M., Meng C.K., Wong S.F., Whitehouse C.M., *Science* **1989** Oct 6, *246*(4926), 64–71.

Electrospray Mass Spectrometry of Poly(ethylene glycols) with Molecular Weights up to Five Million Nohmi T., Fenn J.B., **1992**.

Three-dimensional Deconvolution of Multiply Charged Spectra Labowsky, M.J., Whitehouse C.M., Fenn, J.B., *Rapid Commun. Mass Spectrom.* **1993,** *7*, 71.

Ion Formation from Charged Droplets: Roles of Geometry, Energy and Time Fenn, J.B. *J. Am. Soc. Mass Spectrom.* **1993**, *4*, 524.

Mass Spectrometry with Ion Sources Operating at Atmospheric Pressure Bruins, A.P. *Mass Spectrom. Rev.* **1991**, *10*, 53–77.

Liquid Chromatography-Mass Spectrometry. General Principles and Instrumentation, Niessen, W.M.A., Tinke, A.P. *J. Chromatogr. A* **1995**, *703*, 37–57.

Capillary Liquid Chromatography Mass Spectrometry Tomer, K.B., Moseley, M.A., Deterding, L.J., Parker, C.E. *Mass Spectrom. Rev.* **1994**, *13*, 431–457.

Liquid Chromatography-Mass Spectrometry and Related Techniques via Atmospheric Pressure Ionization, Wachs, T., Conboy, J.C., Garicia, F., Henion, J.D. *J. Chromatogr. Sci.* **1991**, *29*, 357–366.

Internet addresses of interest for LC-MS coupling

http://www.spectroscopynow.com
http://www.asms.org
http://www.dgms.de
http://www.lcms.com
http://www.chemistry.gatech.edu/stms
http://www.ionsource.com
http://masspec.scripps.edu/information/history/

2.2 Micro- and Nano-LC

Jürgen Maier-Rosenkranz

A short introduction

The use of columns with smaller inner diameters is becoming increasingly important – a development enhanced by the need for better sensitivity and the use of mass spectrometers. These applications work with smaller flow rates, which put extra demands on the HPLC systems, especially pumps, detectors and autosamplers as well as the connecting parts.

Although a wide range of good quality micro- and nano-LC columns have been commercially available for a long time, progress in developing the other components has been somewhat slower. Nowadays, however, there is a range of suitable devices to choose from – depending on the volume of the system.

It seems therefore a good idea to classify the different ranges of the analytic HPLC systems according to flow rates.

A user wanting to use columns with an ID <3 mm in his HPLC system will first of all have to find out if the system meets the requirements for working with such small volumes.

Therefore, before you start working with columns with ID ≤2 mm, you must establish if your system is suitable for the specific purpose, which includes performing an efficiency test and testing the accuracy of the flow rate and gradient.

A table to record all the volumes measured in your system (all tubing, capillaries, devices) can prove extremely useful. Based on the calculation of limit values, you can establish the column dimensions for the system in question.

If these criteria are met there should be no technical problems working with HPLC columns with an ID of 50 µm.

Table 2-1.

	Inner diameter of columns	**Flow rate**
Analytical LC	4.6 mm	1325 µl/min
	4.0 mm	1000 µl/min
	3.0 mm	563 µl/min
Microbore-LC	2.0 mm	250 µl/min
	1.6 mm	160 µl/min
	1.0 mm	62.5 µl/min
Micro-LC	800 µm	40.0 µl/min
	500 µm	15.6 µl/min
	300 µm	5.6 µl/min
Nano-LC	180 µm	2.03 µl/min
	150 µm	1.41 µl/min
	100 µm	0.63 µl/min
	75 µm	0.35 µl/min
	50 µm	0.16 µl/min

Tip No. 82 — Lower efficiency – plate number too low

Part 1: Effects of dead volume in the connecting parts. Which column diameter should be used with which capillary diameter?

The capillary connections from the injector to the column and from the column to the detector affect directly the plate number of the peaks, and thereby also the resolution. In order to make the best use of the efficiency of the column, the following values should not be exceeded:

For columns with an ID of 1 and 2 mm: column volume/50 ≥ capillary volume
For columns with an ID ≤ 800 µm: column volume/30 ≥ capillary volume

This cannot be applied consistently to columns with an ID ≤ 500 µm, because the danger of blocking the capillaries with an ID < 50 µm increases enormously.

The recommendations shown in the Table 82-1 are a sensible compromise between efficiency, resolution and the risk of clogging-up.

Table 82-1.

Column		Capillary	
ID	Length	ID	Length
2 mm	≥ 100	≤ 170 µm	< 15 cm
2 mm	< 100	≤ 130 µm	< 10 cm
1 mm	≥ 100	≤ 130 µm	< 10 cm
1 mm	< 100	100 µm	< 10 cm
800 µm	all	75 µm	< 30 cm
500 µm	all	75 µm	< 30 cm
300 µm	all	75 µm	< 30 cm
200 µm	all	50 µm	< 30 cm
180 µm	all	30 µm	< 30 cm
150 µm	all	30 µm	< 30 cm
100 µm	all	25 µm	< 30 cm
75 µm	all	25 µm	< 20 cm
50 µm	all	20 µm	< 20 cm

Tip No. 83

Lower efficiency – plate number too low

Part 2: Effects of injection amount and injection volume

A large injection volume, overloading and using the wrong solvent can decrease the plate number considerably and thus compromise the entire result of a separation. The following conditions should be observed:

RP chromatography under isocratic conditions:
- Injection volume: The injection volume should not exceed 10% of the flow rate per minute. Thus the loading time for the sample is <6 s.
- Solvent: The amount of organic modifier (ACN, MeOH) should not exceed the amount contained in the mobile phase. The larger the injecting volume, the lower the organic proportion must be.
- Sample amount: Obviously, the amount of sample depends strongly on the type of substance. The following rule of thumb is very helpful – approx. 0.01 µg substance per µL column volume can be loaded.

Table 83-1 illustrates these recommendations.

To avoid these problems in micro- and nano-LC you usually work with gradients.

In a first step, the sample is focused at the column head under RP conditions with a high proportion of water and then eluted in a linear gradient. This makes it possible to work with sample volumes of between 1 and 50 µL on capillary columns.

Table 83-1.

Length of column (mm)	ID (mm)	Volume (µL)	Flow rate (µL/min)	Max. injection volume (µL)	Max. load (µg)
125	2	393	250	25	3.93
50	2	157	250	25	1.57
125	1	98	50	5	0.98
50	1	39	50	5	0.39
125	0.800	63	32	3.2	0.63
50	0.800	25	32	3.2	0.25
125	0.500	25	12	1.2	0.25
50	0.500	9.81	12	1.2	0.10
125	0.300	8.83	5	0.5	0.09
50	0.300	3.53	5	0.5	0.035
125	0.180	3.18	1.6	0.16	0.032
50	0.180	1.27	1.6	0.16	0.013
125	0.100	0.98	0.5	0.05	0.010
50	0.100	0.39	0.5	0.05	0.004
125	0.075	0.55	0.3	0.03	0.006
50	0.075	0.22	0.3	0.03	0.002

Tip No. 84 Lower efficiency – plate number too low

Part 3: Impact of flow cell (UV, fluorescence, radio detection)

The flow cell volume has a considerable impact on the plate number of the peaks in question, which is why it is crucial to use the correct size of flow cell for the separation in hand.

A flow cell is characterized by three parameters:
- Volume
- Path length
- Signal-to-noise ratio

Volumes and path length are the standard characteristic data indicated in the manual or data sheet, while it is left to the user to work out the signal-to-noise ratios to compare flow cells.

Table 84-1 indicates which cell volume is appropriate for which column ID.

At the maximum flow cell volume, 10 times the peak volume is flushed through, and the plate number is only just a third. In order to reach the maximum plate number, the peak volume must be 50 times larger than the flow cell volume.

These correlations are also shown in Figure 84-1. Plate number and resolution have both increased considerably, and only now can some of the peaks be integrated. Here the loss in sensitivity by about 50% hardly matters (see also Tip No. 85).

Table 84-1.

Flow cells for broad peaks (here 30 s base width)					Flow cells for sharp peaks (here 6 s base width)				
ID (mm)	Flow rate (µL/min)	Peak volume (µL)	Flow cell max. (µL)	Flow cell opt. (µL)	ID (mm)	Flow rate (µL/min)	Peak volume (µL)	Flow cell max. (µL)	Flow cell opt. (µL)
4.6	1000	500	50.0	10.0	4.6	1000	100	10.0	2.0
4.0	750	375	37.5	7.5	4.0	750	75	7.5	1.5
3.0	500	250	25.0	5.0	3.0	500	50	5.0	1.0
2.0	250	125	12.5	2.5	2.0	250	25	2.5	0.5
1.0	50	25	2.5	0.5	1.0	50	5	0.5	0.1
0.800	32.00	16	1.600	0.320	0.800	32.00	3	0.320	0.064
0.500	12.50	6.25	0.625	0.125	0.500	12.50	1.25	0.125	0.025
0.300	4.50	2.25	0.225	0.045	0.300	4.50	0.45	0.045	0.009
0.200	2.00	1.00	0.100	0.020	0.200	2.00	0.20	0.020	0.004
0.100	0.50	0.25	0.025	0.005	0.100	0.50	0.05	0.005	0.001

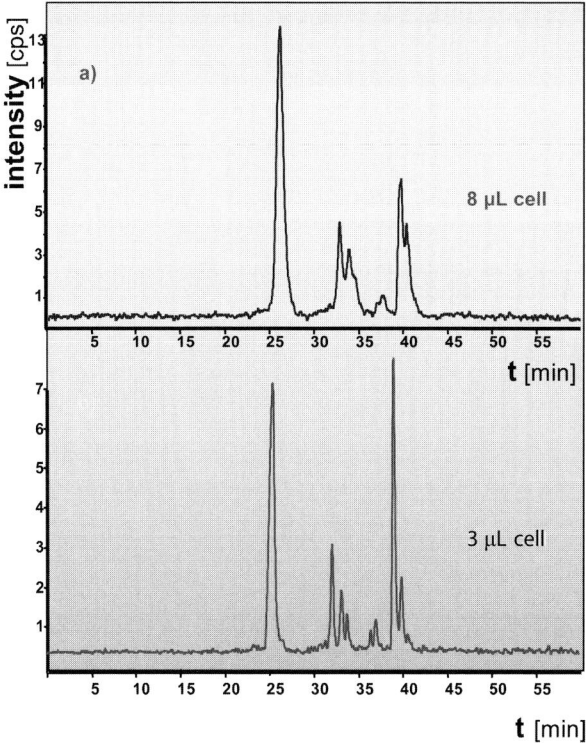

Figure 84-1. Comparison of two cells, one with 8 and the other with 3 µl volume. M. Breyer, M. Twele, K. Schmeer, P. Földi. **Stationary Phase:** Nucleosil 100 C 18 HD, **Column diameter:** 125×1 mm 3 µm, **Flow rate:** 37.5 µl/min, **Eluent:** A: 10 mM NH_4Ac pH 3.0, B: ACN, **Flow cell:** a) 8 µl, b) 3 µl, **Sample:** Drug+Metabolite, **Injection:** 40 µl.

Tip No. 85 — No gain in sensitivity: flow cell – path length – S/N

Which cell/column combinations achieve the highest sensitivity can be worked out with the help of Table 85-1. It shows very clearly that sensitivity increases with the decreasing column ID and increases with greater path length.

No account is taken of the signal-to-noise ratio in this estimate, but it can be generally said that smaller cells perform less well. The table shows the respective theoretical optimums.

With flow cells <1 µL with path lengths >3 mm the signal-to-noise ratio is often more than 10 times worse than with large volume cells.

Table 85-1.

ID (mm)	Flow rate (µL/min)	Flow cell volume (µL)	Flow cell path length (mm)	Increase in sensitivity
4.6	1000	15	10	1.0
4	750	15	10	1.3
3	500	*15	10	2.4
		9	6	1.4
2	250	*15	10	5.3
		*9	6	3.2
		1.2	3	1.6
1	50	*15	10	21.2
		*9	6	12.7
		1.2	3	6.3
0.8	32	*15	10	33.1
		*9	6	19.8
		1.2	3	9.9
		0.035	0.4	1.3

Tip No. 86 — Fused silica and PEEK capillary connections

Insufficient capillary connections are frequently to blame for a loss in resolution. In order to achieve a good connection and avoid a dead volume, the end of the capillary must be smooth and cut at right angles. It is best to use tools with a knife rotating round the capillary. To finish off, use a grinding block, which is cheap and easy to use and suitable for all materials (steel, PEEK, fused silica). Using a PEEK finger-tight fitting, the capillary can be locked into the block so that it sticks out by just a few tenths of a millimetre. Then put very fine abrasive paper (400–800) on a smooth flat surface and run the block over it in circular movements. To finish, flush the capillary thoroughly.

Before installing capillaries in your apparatus it is vital to check their ends thoroughly. Use a magnifying glass with 10–20 fold magnification.

For stabilizing and sealing fused silica or thin PEEK capillaries, special sleeves with an outer diameter of $1/16^{th}$ inch (=1.6 mm) that fit on all standard HPLC connections can be used. There are even thinner $1/32^{nd}$ inch (=0.8 mm) ones available for micro finger-tight fittings, which are easier to handle and help seal the connections.

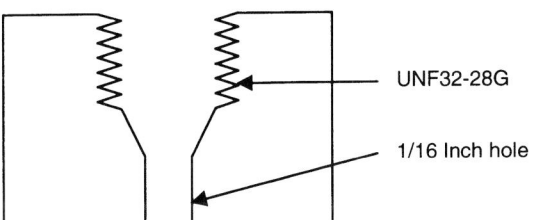

Figure 86-1. Cross-section of grinding book.

Tip No. 87 — Fast sample loading due to column switching

As described in Tip No. 83, the loading of a sample is crucial. In order to avoid a loss in resolution, the sample is usually enriched at the head of the column. There are, however, two problems that arise in the process. In samples of a volume between 5 and 25 µL the loading time is in the range of a few minutes, and with a 300 µm ID column and a flow rate of 5 µL min^{-1} it takes 1–5 min to load the sample. In addition, the non-retained part of the sample, such as polar components of the matrix, is pumped through the whole column.

These problems can be avoided by using a guard column switching system (Figure 87-1). The sample is enriched on a guard column, and all non-retained substances end up in the waste container. The guard column, which has a larger ID than the main column, should be as short as possible. The larger ID permits the use of higher flow rates for sample loading, and the risk of the pressure increase in the system by impurities is clearly smaller. The shortness of the column keeps volume and pressure low.

Ideally, the ID of the guard column should be approximately twice that of the separation column. The length is between 5 and 10 mm. Suitable combinations of pre-column and separation column are shown in Table 87-1.

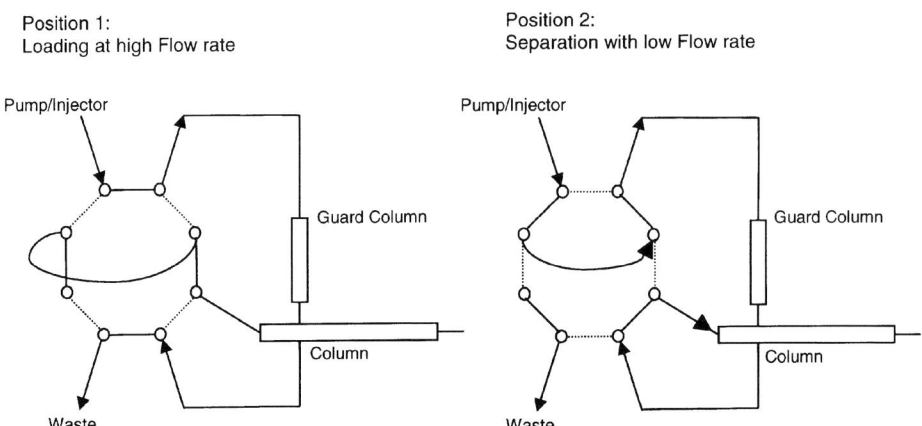

Figure 87-1. Guard column switching: 2-position-8-path-valve.

Table 87-1. Columns with suitable guard columns for sample enrichment

Separation column		Guard column			
ID (mm)	Flow rate (µL min^{-1})	ID (mm)	Length (mm)	Flow rate (µL min^{-1})	Volume (µL)
2.000	250	4	5	2000	62.80
2.000	250	3	10	1200	70.65
1.000	50	2	5	500	15.70
1.000	50	1.6	10	300	20.10
0.800	32	1.6	5	300	10.05
0.800	32	1	10	120	7.85
0.500	12	1	5	120	3.93
0.500	12	0.8	10	75	5.02
0.300	5	0.5	5	30	0.981
0.300	5	0.5	10	30	1.963
0.180	1.6	0.3	5	10	0.353
0.180	1.6	0.25	10	7	0.491
0.100	0.5	0.17	5	4	0.113
0.100	0.5	0.13	10	2.5	0.133
0.075	0.3	0.13	5	2.5	0.066
0.075	0.3	0.13	10	2.5	0.133

Tip No. 88 — Injection system: full loop injection, partial loop fill injection, timed programmed injection, direct injection

As described in Tips Nos. 83 and 87, the loading of the sample has a substantial effect on the quality of the result, which is why I would like to describe the most common techniques from a micro- and nano-LC perspective.

Full loop injection

This injection method usually provides good accuracy and reproducibility. If the sample is concentrated onto the column or guard column after the injection, normal analytical injectors can be used, as long as the flow rate is kept high enough to flush through the total volume of the injector within 30 s. For lower flow rates specific nano-injectors should be used, which come in two types:

1. The loop is cut into the rotor (60 or 200 nL). The injection volume remains constant with this type.
2. Micro-6-way-2-position-valve. Fused silica capillaries serve as a loop.
   The use of different capillaries makes injecting a range of volumes possible.

Partial loop injection

This method is only practicable in combination with sample enrichment on the guard column or on the column head. Here, too, the volume is crucial.

Nano-injectors cannot be used for partial loop injection.

Timed programmed injection

This is a special form of full or partial loop injection. It involves partially or completely filling the loop and using a timing control to regulate the flow rate towards the column. Thus volumes of from 10 nL to 10 µL can be injected accurately and flexibly.

Example:
Flow rate 5 µL min^{-1}, the loop is connected for 6 s – injected volume 0.5 µL.

Direct injection

With this variant the looped injection needle is always in the flow towards the column. This is not suitable for flow rates <15 µL because the gradient delay volume would become too large.

Tip No. 89

Protecting the system: cleaning-up guard column, saturation column

With a smaller column ID, the risk of higher pressure and blockage increases. This may compromise the stability of the system, and the columns must be replaced frequently.

The best protection for the system is a cleaning-up guard column. This column – not to be confused with a guard column – is packed with coarse RP silica gel (20–50 µm) and placed between the pump and injector. Thus all impurities (abrasives from the pump, microorganisms, etc.) from the mobile phase are filtered out, while at the same time the mobile phase is saturated with silicic acids. With high-pressure gradient systems it can be also used as a mixing chamber.

The volume of this cleaning-up guard column must be adapted to the respective flow rates to keep the gradient delay volume within reasonable limits. As the column must be flushed through within <30 s, the dead volume of the cleaning-up guard column must not exceed 50% of the flow rate per minute. Table 89-1 gives an overview of the dead volumes to be expected.

Table 89-1. Dead volume of cleaning-up guard columns

Guard column			Guard column		
ID (mm)	Length (mm)	V_0 (µL)	ID (mm)	Length (mm)	V_0 (µL)
4	5	44.0	4	10	87.9
3	5	24.7	3	10	49.5
2	5	11.0	2	10	22.0
1.6	5	7.03	1.6	10	14.1
1	5	2.75	1	10	5.50
0.8	5	1.76	0.8	10	3.52
0.5	5	0.69	0.5	10	1.37
0.3	5	0.25	0.3	10	0.49
0.25	5	0.17	0.25	10	0.34
0.17	5	0.08	0.17	10	0.16
0.13	5	0.05	0.13	10	0.09

Tip No. 90 — Retention time shift: gradient delay volume, mixing chamber volume, gradient accuracy

A frequent problem in micro- and nano-LC is the instability of the retention times. There are considerable differences from one injection to another, which makes the evaluation of the chromatograms rather difficult. The reason lies in too large a gradient delay volume in the system.

The gradient delay volume is the sum of all volumes from the point in the HPLC system where the gradient is produced (mixing chamber, T-fitting) up to the column head.

The influence of the mixing chamber volume on the gradient profile is shown in Figure 90-1.

After 1 min flushing time (5 µL mixing chamber volume and 5 µL min^{-1} flow rate), 100% B is not reached with a short gradient, even though this level has been held for 2 min. The effect on the wash-out-characteristic, however, is much more dramatic when the gradient is changed back to the starting concentration. It takes 5 min until the correct concentration at the column head is achieved, and at least a further

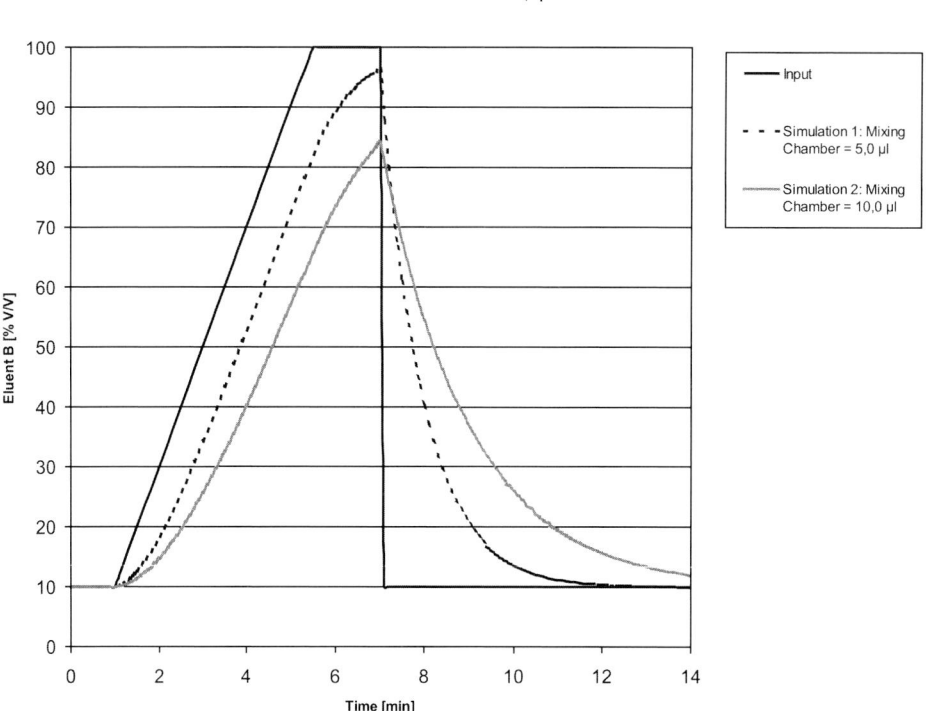

Figure 90-1. Impact of the mixing chamber volume on the gradient profile.

5 min for the column. The total equilibration time is thus at least twice the gradient time. Therefore, the next sample has often been injected although the column has not yet equilibrated, and this causes varying retention times.

Therefore, it is recommended that the gradient delay volume is kept small so that it can be flushed within 1 min. Subsequently, the time needed to reach a safe and constant equilibration should be determined.

The same applies even if you are working with a splitter. The primary flow rate must also flush the primary gradient delay volume within 1 min (see also Tip No. 91).

Tip No. 91 Transferability – downscaling: correct gradients

When transferring a method from a larger to a smaller column ID, the shape of the gradient is crucial. In most commercially available HPLC systems, however, the real gradient is a parameter that is dependent on the flow rate.

Example:
According to the manufacturer, a system can be appropriate for applications for columns with an ID of from 2 to 4.6 mm. The gradient delay volume amounts to 1 mL. If one works with a 4 mm ID column with a flow rate of 1 mL min^{-1}, the gradient delay time is 1 to 2 min. If working with a 2 mm ID column with 0.25 mL min^{-1}, the gradient delay time is >5 min, the gradient form changes considerably, and the discrepancy between the programmed and the real gradient is simply unacceptable.

You must therefore work out precisely for what applications it is suitable.

Recommended test:
Gradient:
A = water; B = acetonitrile with 0.1% benzyl alcohol.
Detection:
UV 254 nm
Column:
Instead of a column, a thin capillary is used in which the pressure rises to at least 20 bar.

An example is shown in Figure 91-1.

Table 91-1. Gradient for acetonitrile with 0.1% benzyl alcohol

Time (min)	B (%)
00.00	0
02.00	0
02.01	20
04.00	20
04.01	40
06.00	40
06.01	60
08.00	60
08.01	80
10.00	80
10.01	100
20.00	100

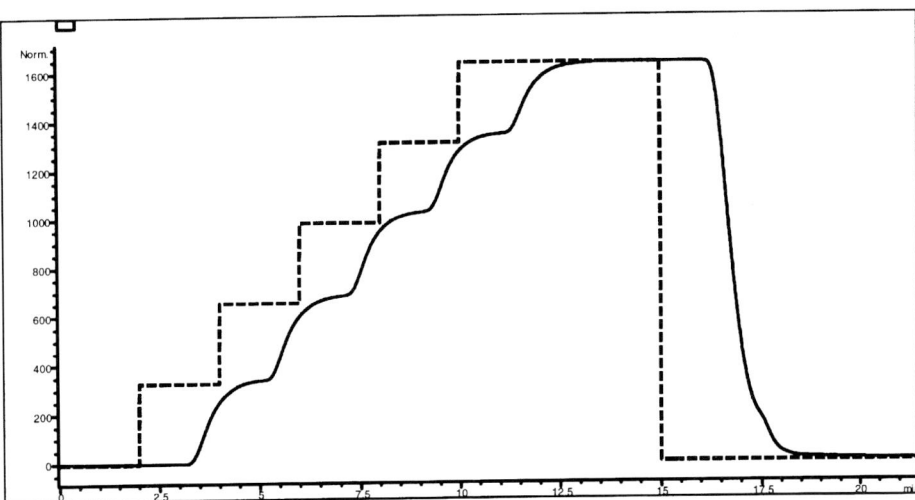

Figure 91-1. Real gradient profile

The response time (first point showing an increased signal after the change in concentration) is:

3.2 min – 2.0 min = 1.2 min

The rising edge time (reaching the 1st concentration plateau) is:

5.0 min – 2.0 min = 3.0 min

The simplest way to achieve comparable and accurate gradients is by defining accuracy limits for the gradients.

Option 1: Limit the values for response and rising time:
        Response time: <1 min, with short gradients (<10 min): <0.5 min
        Rising time: <2 min, with short gradients (<10 min): <1.5 min

Option 2: Determination of a flow rate limit
        The gradient profile described above is produced with different flow rates over the full flow range of the system. The flow rate limit is defined as the flow rate that leads to a discrepancy in the respective graphs.

References to Micro- and Nano-LC

Versele, M., Dewaele, C.J. *High Resolut. Chromatogr. Commun.* **1987**, *10*, 280.
Novotny, M. *Anal. Chem.* **1988**, *60*, 500.
Ishii, D., Takeuchi, T. *Trends Anal. Chem.* **1990**, *9*, 152.
Grom, E., Maier-Rosenkranz, J., Földi, P. *Laborpraxis* **1996/3**, 16–22.
van Straten, M.A., Vermeer, E.A., Claessens, H.A., *LC-GC Int.* **1996**, *1*, 42.
Vissers, J.P.C., Claessens, H.A., Cramers, C.A. *J. Chromatogr. A* **1997**, *1*, 779.
Vissers, J.P.C. *J. Chromatogr. A* **1999**, *856*, 117–143.
Maier-Rosenkranz, J., *Laborpraxis* **2000/3**, *24*, 18–23.
Breyer, M., Twele, K., Schmeer, K., Földi, P. *Laborpraxis* **2001**, *09*, 18–22.
Palmer, C., Remcho, V. *Anal. Bioanal. Chem.* **2002**, *372*, 35.
Kientz, C.E., Langenberg, J.P., De Jong, G.J., Brinkman, U.A.T. *J. High Resolut. Chromatogr.* **1991**, *14*, 460.
Klein, C., Földi, P. *Laborpraxis* **2002**, *02*, 34–38.
Prüß, A., Kempter, C., Gysler, J., Jira, T. *J. Chromatogr. A* **2003**, *1016*, 129–141.

2.3 Quantification

2.3.1 Practical aspects of quantification in HPLC

Stavros Kromidas

The first part of this chapter (2.3.1) will highlight just a few points concerning quantitative analysis with the emphasis on practical issues.

- Peak area or peak height?
- What factors have an impact on the peak area?
- Formulae that apply to individual quantification methods
- Examples of calculations

The second part (2.3.2) will deal extensively with various aspects of quantitative analysis, including the not-so-trivial question of weighted regression.

2.3.1.1 Peak area or peak height?

Nearly all regulations say explicitly that quantification can be carried out using the peak area as well as peak height. The fact is, however, that, apart from a few exceptions, it is the peak area that is normally referred to for quantification. This is perfectly all right, except for the following cases:

1. Insufficiently resolved peaks.
2. Peaks near the limit of quantification, especially if there is tailing or a drift on the baseline.
3. The short-term consistency of the flow rate is unsatisfactory and no remedial or other such measures improve the situation.

Case 1:

If the resolution is insufficient (peaks close together, shoulder peaks, etc.), quantification by peak area is always somewhat risky. Whether you use the tangent method or draw a perpendicular line, the problem will not go away: no software can really work out the shape of an un-resolved peak. It simply does not "know" if the peak is symmetrical or if it is tailing. Thus, quantification by peak area can lead to considerable errors, given that the peak shape of the analyte from the standard solution could differ from the shape it has when it is close to another peak. By contrast, the peak height is fairly independent of the degree of peak overlap. In any case, the quantification by peak height seems to be the lesser of two evils – unless you can find an easy way of improving chromatographic resolution.

Case 2:

Even in manual integration it is often difficult to find objective criteria to define the beginning and the end of small tailing peaks. These peaks need to be looked at individually, and there may be inconsistencies of between 5 and 10% in repeat measurements. In automatic integration, inconsistencies may arise from an incorrect setting of the integration limit. For small peaks, this could amount to an error of several percent, see Figure 2-2.

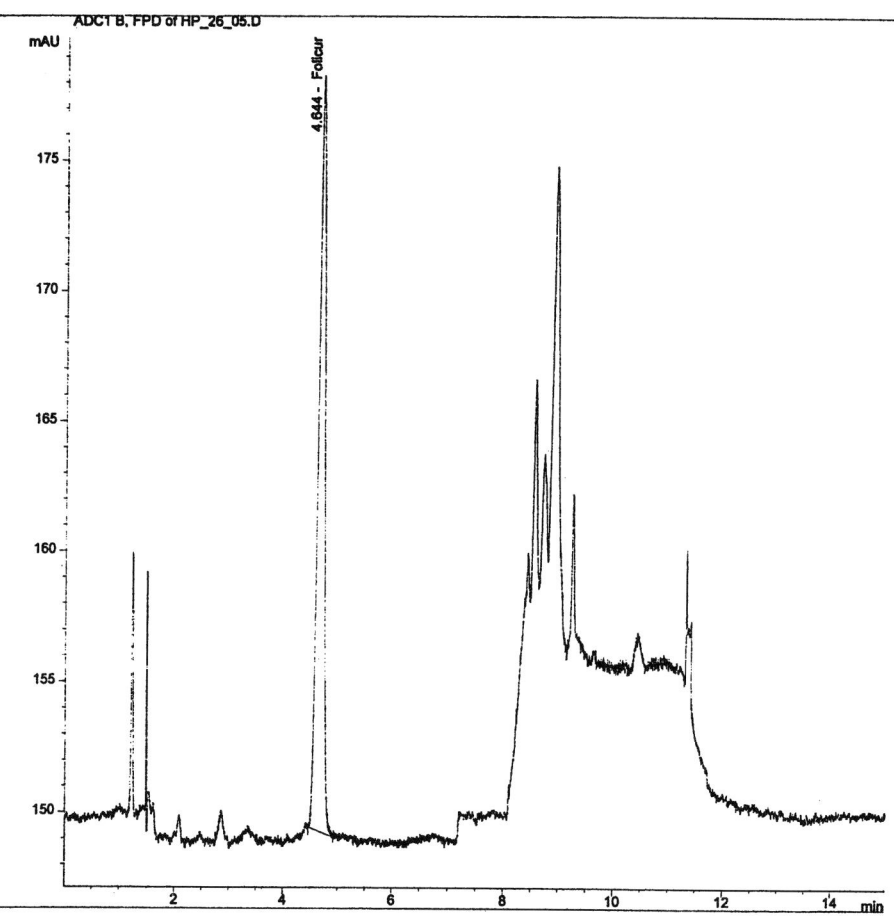

Figure 2-2. Concerning individual (subjective) integration boundaries: Depending on their choice, the peak at 4.6 min can vary by up to 7% in area.

Case 3:

Short-term flow fluctuations in the pump (we are talking about fluctuations during the elution of the peaks) mainly influence the peak area, but the peak height is affected to a much lesser extent. If the problem persists even after the pump has been repaired, evaluation by peak height is the preferred option. Finally, if retention times are unstable, the quantification becomes more reliable if the calculation is based on the product of retention time and peak height. This very effective trick for reliable quantification in isocratic analysis is not known to many people.

2.3.1.2 What factors have an impact on the peak area?

All parameters that can affect peak area or height have an impact on quantification. In this section, we are not including such familiar parameters as injection volume or wavelength, but put the emphasis on those parameter settings that are more easily overlooked in daily routine – see also the more extensive discussion in part 2 of this chapter (2.3.2). We refer exclusively to quantification via peak area, as this remains the prevalent method:

- The greater the number of data points (sampling time), the more accurately the peak will be represented, which is particularly noticeable in early narrow peaks.
- The smaller the time constant or think "bunching factor/bunching rate", the narrower the peaks, and the more accurate the integration will be. See Tip No. 68.
- If, on the other hand, too large a peak width is chosen, narrow peaks could be overlooked or, in early peaks, the integration may be inaccurate because the peaks appear broader and resolution would seem to be deteriorating.

The peak area, or, to be more precise, the determination of the peak area can, of course, also be affected by noise or, more indirectly, by parts of the instrument malfunctioning or by its immediate environment.

Example:
Insufficient grounding of the PC or the detector, insufficient protection against surges, unshielded detector cables, a faulty column oven, slow AD converters, etc., can all cause a distortion of the peaks. A high frequency noise level resulting from the same causes may make the beginning and the end of the integration difficult to define.

The difficulties discussed here are particularly noticeable in early-eluting small peaks. The evaluation of relatively large peaks, quite common in assay analysis, is less problematic.

When the quality of the column packing declines, contaminations that elute in the immediate proximity of the main peak could remain undetected, and this excess area could be added to the main peak area and result in an error of several percent.

Finally, some compound-specific causes should be mentioned that could result in a decrease or change in peak area, such as an unstable sample, irreversible adsorption of the sample to steel or other surfaces in the instrument, pH dependency of UV absorption.

2.3.1.3 Formulae and short statements or comments with respect to the quantification methods

1. External standard method

$$c_x = \frac{c_{st}}{A_{st}} \cdot A_x = R_{fc} \cdot A_x \ (\text{mg mL}^{-1})$$

$$m_x = \frac{W_{st}}{A_{st}} \cdot A_x = R_{fm} \cdot A_x \ (\text{mg})$$

$$m_{x\%} = \frac{R_{fm} \cdot A_x}{w} \cdot 100\%$$

$$W_{Akt} = c_x \cdot V_{inj}$$

c_x = concentration of the (unknown) analyte
c_{st} = concentration of the calibration substance (e.g. in mg/mL)
A_{st} = area of the calibration substance

A_x = area of the analyte
$R_{fc,m}$ = response factor (concentration or amount)
m_x = mass of analyte in weighed sample
w_{st} = 100% calibration substance weighted in mg
$m_{x\%}$ = mass of the analyte in the sample in %
W = weighted sample in mg
W_{Akt} = injected mass of the analyte
V_{inj} = injection volume

- The main advantage of the external standard method is that the standard is usually the same compound as the compound you are looking for. Thus, there is no discrepancy in the response factor (relationship between signal–concentration) [5].
- If the method is robust and the instrument is in impeccable working order (high quality and stable standard, easy sample preparation, constant injection volume, stable chromatographic conditions, etc.), the external method is the simplest and should be the method of choice.

2. Internal standard method

$$c_x = c_{komp} \cdot \frac{A_{ist}}{A_{komp}} \cdot \frac{A_x}{A'_{ist}}$$

c_x = unknown concentration of the analyte in the sample
c_{komp} = concentration of the analyte in calibration
A_{ist} = area of internal standard in calibration
A_{komp} = area of analyte in calibration
A_x = measured area of analyte
A'_{ist} = measured area of the internal standard

- The internal standard is recommended for all those cases where something untoward may happen, such as:
  - Several complex steps in the sample preparation that could lead to sample losses, e.g., extraction, precipitation or derivatization
  - Unstable sample
  - Losses during the injection process

- Slightly simplified, the difference between the internal and external standard method could be described as follows: for the external standard method, there is a relationship between peak areas and concentrations, whereas for the internal standard method there is a relationship between area ratios and the corresponding concentrations.

What are the properties of a good internal standard?

- Its chemical and physical properties should resemble those of the component to be analysed as far as possible. However, the chance of it being in the sample must be excluded, even as a possible future event.
- It separates well from other compounds and it elutes near the relevant peak in the chromatogram. Sometimes two internal standards are needed to meet these requirements, one that elutes early and one that elutes late in the chromatogram.

- If the response factor is similar to the compound of interest, the internal standard should yield an area of similar size.
- The internal standard should be stable and clean and not react with other sample components or the matrix.
- The suitability of the internal standard should be statistically proven using the external standard method.

3. **100% method or: internal normalization; normalization using response factors; normalized area method; normalization to 100%**

    If the response factors of all components are equal, the following equation applies:

    $$m_i = \frac{A_i}{\sum A_i} \cdot 100$$

    If the response factors differ, the following formula applies:

    $$m_i = \frac{A_i \cdot R_{fi}}{\sum R_{fn} \cdot A_n} \cdot 100$$

    A_i = area of the component i
    $\sum A_i$ = sum of all areas
    R_{fi} = response factor of the component i
    $\sum R_{fn} \cdot A_n$ = sum over all sample components individual response factors multiplied with their respective peak areas

- This evaluation method is popular in routine analyses. As only proportions are being determined (area percentages), the injection volume does not matter in this calculation method. The measured peak areas turn up as numerators as well as denominators, which means that the quotient does not change.
- The 100% method has its place in GC where response factors tend to be similar. In HPLC UV detection it can be quite labour-intensive as sample components often have a wide variety of extinction coefficients (!). Establishing an internal normalization for refractive index detection makes sense and is easy to do. In order to use the 100% method correctly you must be able to detect all the substances in a sample.

4. **Standard additions method**

    $$x_i = \frac{\Delta x_i}{\dfrac{\Delta A_i}{A_i} \cdot \dfrac{A_{st1}}{A_{st2}} - 1}$$

    x_i = original amount of analyte in the sample
    Δx_i = sample quantity x_i, added quantity of i
    ΔA_i = peak area of analyte after adding Δx_i
    A_i = peak area of analyte before adding Δx_i
    A_{st1} = peak area of internal standard before adding Δx_i
    A_{st2} = peak area of internal standard after adding Δx_i

    This method is used when a blank sample (sample that does not contain the relevant compound) is not available or if the matrix must remain unaltered, e.g., if the sensitivity of a detector could be affected by changes in the matrix.

2.3.1.4 Examples with actual figures

The following section gives you the opportunity to apply the various calculation methods in a number of simple examples. All you need is a calculator. You will find the answers on page 253.

Most of these examples were kindly provided by Hans-Joachim Kuss.

Example 1 (external standard method)

A calibration solution with a concentration of 12 mg L^{-1} yields a peak area of 6000 area units. The peak area of a sample solution containing the same analyte is 8000 area units.

- What is the concentration of the analyte in the sample?
- What is the absolute mass of the analyte in the sample if 20 µL are injected?

Example 2 (internal standard method)

In a calibration, injecting an analyte solution at a concentration of 8 mg L^{-1} yields a peak area of 8000 area units. The peak area of the internal standard is 4000 area units. Injecting the sample with the same internal standard concentration results in an internal standard area of 4200 area units and an area of the unknown analyte concentration of 5200 area units.

- What is the concentration of the analyte in the sample?
- What is the absolute mass of the compound in the sample if 10 µL are injected?

Example 3 (addition method, external and internal standard)

The concentration of a pollutant in a sample taken from a landfill site is to be determined. In a first step, a compound is added to the sample to provide an internal standard. The injection of the sample yields the following values:

Area of the analyte (pollutant)	127210 area units
Area of the internal standard	174832 area units

Then 200 mg of the pollutant are added to the sample. A subsequent injection of the sample yields the following values:

Area of the analyte after adding 200 mg	213115 area units
Area of the internal standard	172703 area units

Please calculate the concentration of the pollutant in the original sample using the external and internal standard method.

Example 4 (external and internal standard)

Objective: The metabolites A and B in human urine are to be determined by HPLC after the sample has been adequately prepared.

1. Previous experience in similar projects has raised the suspicion that irreversible adsorption onto the injection block may occur when these substances are injected. We are therefore looking into the possibility of making up for potential injection errors by using the internal standard method. A solution is prepared that contains the pure compounds A and B at a concentration of 1 mg L^{-1} as well as substance P as an internal standard. 50 µL are injected three times. The peak areas that were achieved are given below.

Table 2-2. Peak areas for 50 μL injections of A, B and P

Injection	A (area units)	B (area units)	P (area units)
1.	66 123	68 234	128 345
2.	70 123	72 234	134 345
3.	68 123	70 234	131 345

- What was the absolute mass of pure substance injected?
- Can the use of the internal standard make up for a potential injection error – in other words can the internal standard method be recommended in this particular instance?

2. The metabolites A and B are now to be determined within the real matrix. For this, 50 ng of A, B and P are added to blank urine. The sample is prepared, including extraction and filtration before it undergoes chromatography. These are the resulting areas – A: 49123, B: 51234 and P: 98345
   - What is the amount extracted? Please correlate the mean values of this "real" measurement to the mean values of a standard solution that has been chromatographed without any prior sample preparation – A: 68123, B: 70234, P: 131345.

3. You now have a urine sample with an unknown amount of A and B. 50 ng of P are added as the internal standard. The injection of the sample yields the following areas (note that in real life these mean values would be the results of perhaps six independent tests) – A: 23123, B: 49234, P: 103345.
   - What quantities of A and B are found using the external and the internal standard method? The sample is of course prepared as described as above. Please refer to the values of the standard that have been injected after the preparation of the sample – A: 49123, B: 51234, P: 98345.

Example 5 (spiking, external and internal standard)

An alkaloid (substance A), present in small amounts in a plant extract, can only be detected after several complex sample preparation steps. The procedure is prone to inconsistencies, which is why a standard additions analysis and an internal standard are used: 200 ng of internal standard P are added to the sample containing the unknown endogenous amount of substance A. This is then injected, measurement 1. Then 100 ng of substance A – available as a reference material in good quality – is added and another injection is performed, measurement 2. Please calculate the amount of substance A in the original sample using the external and the internal standard method.

Measurement 1:
A (endogenous amount)   662328 (area units)
200 ng P   213277 (area units)

Measurement 2:
A (endogenous amount plus 100 ng)   773241 (area units)
200 ng P   200438 (area units)

2.3.2 Quantification in Chromatography

Hans-Joachim Kuss

2.3.2.1 Optimum separation – correct peak acquisition

The development of a chromatographic method may be troublesome. Perhaps you have to separate a lot of substances, perhaps you need a very complicated sample preparation procedure and possibly you have to measure close to the limit of quantification, which could be specified by a signal-to-noise ratio of 10:1. When the method has been confirmed and the precision and linearity have been tested, you can breathe a sigh of relief because the method can now be used to obtain analysis results. The analysis process is under statistical control and the result is known, including the uncertainty of the result.

But now the struggle for the individual analysis result begins, which we must check by evaluating the chromatogram. Of course, we know the principle of adding variances. Let us assume that we have identified three influencing factors, A, B and C, in our analysis process. These factors have a coefficient of variation (cv) of 0.5%, 1% and 2%, respectively. Addition of the squares of these values and extracting the root leads to an overall cv of 2.3%. This means that A and B have almost no effect. Thus it is obvious that we only need minimise the influence of C to attain a major

> One factor for the estimation of the coefficient of variation (cv) is the student's t value, which characterises the increasing insecurity that is given with less measured points. The degree of freedom (f) is n-1 for a distribution of values, and n-2 for a linear regression, because the two axes "cost" 2 degrees of freedom. For 10 or more observations the factor approximates 2, which is the characteristic value for the normal distribution.
>
> Excel can be used to calculate t values by means of the TINV function for a defined error probability – here 5% = 0.05.
>
> The table and the diagram show that the t values from a linear regression sharply increase for n<5. With n>10, the laboratory effort will be too high.

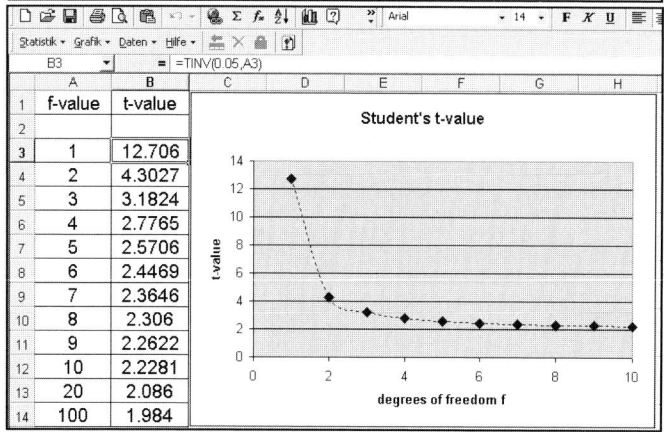

Figure 2-3. What does the Student's *t* value mean?

We analysed one sample containing 2 ng (xl) and one with 20 ng (xu) ten times each. We now want to calculate the two cv's to compare the precision at these two concentrations, which are the lower (xl) and the upper (xu) limits of a working range. We enter the corresponding values of the signals, yl, into cells B1-B10 of an Excel spreadsheet, mark them and name them "yl". Likewise, we enter yu into cells C1-C10, mark them and name them "yu". Use the name box of "B1" to enter "yl" or use <insert> <name>. Enter into B11: |=AVERAGE(yl)| and into B12:|=STDEV(yl)|, to get the two terms necessary to estimate the cv in cell B13. Enter into B13: |=B12*100/B11|, in B14: |=B12/10^0.5| and in B15: |=B14*TINV(0.05,9)|. Mark B11-B15 and expand by one column to the right. Replace yl by yu in C11 and C12.

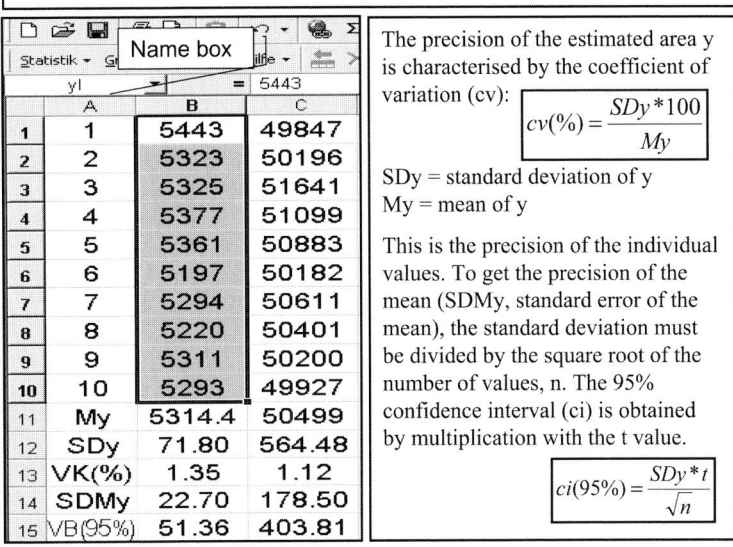

The precision of the estimated area y is characterised by the coefficient of variation (cv):

$$cv(\%) = \frac{SDy*100}{My}$$

SDy = standard deviation of y
My = mean of y

This is the precision of the individual values. To get the precision of the mean (SDMy, standard error of the mean), the standard deviation must be divided by the square root of the number of values, n. The 95% confidence interval (ci) is obtained by multiplication with the t value.

$$ci(95\%) = \frac{SDy*t}{\sqrt{n}}$$

Figure 2-4. How to calculate the coefficient of variation and the confidence interval.

improvement in the method. With these considerations, anything that happens after the detector is often neglected because it can be correctly assumed that the computer is working with high precision. Indeed, you can assume that the error contributed by the integration system is usually very low. However, is that true in all circumstances? Let us take a look at three cases:

1. You have only large, baseline-separated peaks. You see a chromatogram with a straight line due to the baseline, along with narrow individual peaks. Under these ideal conditions, the data system should have no problem. For calibration samples with different concentrations, it will show you an estimated straight line very close to the measured points. This calibration line should be characterised by a correlation coefficient $r=0.999$ and a process variation coefficient [1] of approximately 1%. The intercept should not differ significantly from zero. You can then decide whether to fit a straight line through zero.

If this is not the case, it is possible that the integration conditions are not correctly set. If your slope (threshold) is too high, the data system will find the end of the peak

> Sampling time (ST)
> is a measure of the time interval between two neighbouring data points.
>
> Sampling rate (SR)
> characterises the number of data points per second. In practice, the chromatogram is often measured with a high frequency, e.g. 250 Hz, followed by data bunching to obtain the sampling rate used.
>
> Bunching rate (BR)
> is the number of data points put together to obtain the required sampling rate. This value and the resulting SR are chosen so that a peak is described by 20-30 data points. Therefore, the BR and peak width are related.
>
> Peak width
> is the width of a peak at half height, measured in seconds. This value is deduced from the chromatogram and is used to find the right SR and, for a fixed frequency, the right BR.
>
> Integration suppression
> Normally, the integration of the chromatogram is suppressed from the starting time until after the dead time but before the first peak. In this region, negative peaks can occur that interfere with the integration.
>
> Slope
> If two neighbouring data points are connected, the increase or decrease of the voltage is called the slope. For example, the slope is measured and averaged for one minute and is thus a measure of the level of noise. A beginning or an end of a peak is found when this value is clearly exceeded (or underrun) several times.
>
> Threshold
> Often means the threshold of the slope. It is the limiting value of the slope, and must be exceeded (underrun) for peak identification.

Figure 2-5. Important input data for the integrator.

too early, especially if there is tailing. It is easy to underestimate the change in the integrated area as a result of slight changes in the baseline at the base of the peak. This difference may be several percent, and cannot be neglected [2]. This decrease in the area may be the same for large and small peaks; however, it is more important for the small peaks and can lead to systematic deviations during calibration.

Intentionally wrong adjust the conditions for the integration and take a look at the influence this has.

2. You have small baseline-separated peaks near the limit of quantification. In this case, take a look at the baseline – no longer a straight line – and check for plausibility. Even if the baseline is slightly too high, the area will be reduced. If you have baseline waves near the peak, the area will be increased. Similar considerations as for point 1 apply; however, the calibration will give only $r=0.99$ and a cv of several percent. It can be difficult to set optimum conditions for the data system if you have large as well as small peaks. Another complication is baseline drift caused by the temperature program or the eluent gradient.

3. The advantage of baseline separation is so great, from the quantitative point of view, that it is worth making in a lot of effort to achieve this. Unfortunately, it is not always possible to do this all the time. In reality one has to live with overlapping peaks and shouldered peaks. In this case, you should be aware that large errors are possible, especially with decreasing separations, with increasing tailing and with an increasing difference between the lowest and the highest concentrations. If there is tailing, there is a difference whether the smaller peak lies in front of the larger peak or behind it [3]. The partial overlapping of three peaks is even more difficult, especially if the smallest peak is in the middle. Dyson [4] reported: "There is no integrator solution to this problem; the only solution is improved peak resolution".

To check for these possible errors, one should prepare a calibration curve for a sample with 4 peaks using two different concentration patterns: with the usual method as shown in Table 2-3(a) and also as shown in Table 2-3(b).

Now any unwanted falsifications should be obvious. Such a test could be included in the confirmation of robustness. The above example just demonstrates the underlying principle. Only you know which differences in concentrations are possible and which tolerances are acceptable without affecting the results.

People familiar with daily routine analyses will have had the following experience: the chromatograms of a series of analyses contain a large desired peak, which is followed by a small undesired peak that has not been fully separated. Each of these peaks are integrated separately, but not in every case because only one peak is found in some chromatograms so that the area is the sum of both peaks. The height remains unchanged in such examples.

Low baseline displacements may have a great effect on the area, while the effect on the height is only slight. A peak containing a relatively large shoulder will have an altered area, but only a slightly modified height. Don't believe that an evaluation on the basis of the height is always better than one based on the area. In well-separated chromatograms repeated injection of the same sample usually gives a lower cv if the area is used for the evaluation. The point is: the height is less prone to errors than the area. In practice, this advantage can be crucial. The more complex your chromatogram is and the closer to the limit of quantification you are measuring, the greater the preference you should give to an estimation using the height. But here, similarly to the use of an internal standard, you have to prove the lower variability with realistic samples. For example, collect real analysis samples with a low concentration and measure the pool repeatedly, comparing area and height.

Table 2-3. Preparation of calibration curves

	Peak 1	Peak 2	Peak 3	Peak 4
(a) Concentration pattern 1				
Concentration 1	1	1	1	1
Concentration 2	10	10	10	10
Concentration 3	100	100	100	100
(b) Concentration pattern 2				
Concentration 1	1	100	1	100
Concentration 2	100	1	100	1

2.3.2.2 Understanding the "mind" of the integration system

It is well known that computers only process digital data – only the information 0 or 1, as shown in Table 2-4.

The number 20 is represented by 010100 (16+4) and 45 by 101101 (32+8+4+1). The analogue signal is converted by an analogue-to-digital converter (ADC). An ADC with 20 bit can divide the maximum signal of a UV detector (usually 1 V) into 10^6 parts, that is 1 µV, because 2^{20} corresponds to approximately 1 million. If a very noise-free UV detector is used, this could be insufficient.

An HPLC chromatogram of 30 min is shown in Figure 2-6.

Three main peaks are immediately obvious. The same chromatogram looks even more simple to the integrator, but it is more complicated to "see" the peaks. In the time window in Table 2-5 from 12 to 12.95 min a baseline is shown.

In the time window from 18.5 to 19.45 min (Table 2-6), there is a peak with a maximum at 18.85 min.

The data system fits a parabolic curve to the highest values and estimates the exact retention time and the exact height of the peak. The computer needs very little time to analyse the chromatogram in its own way, but the "thinking" is different to ours and more simple. Excel can convert the entire table into a diagram that looks like our usual chromatogram. The difference between two consecutive points is the slope, or the first derivation with respect to time. The second derivation is found analogously.

Table 2-4. Processing of digital data

Exponent	5	4	3	2	1	0
$2^{exponent}$	32	16	8	4	2	1
20	0	1	0	1	0	0

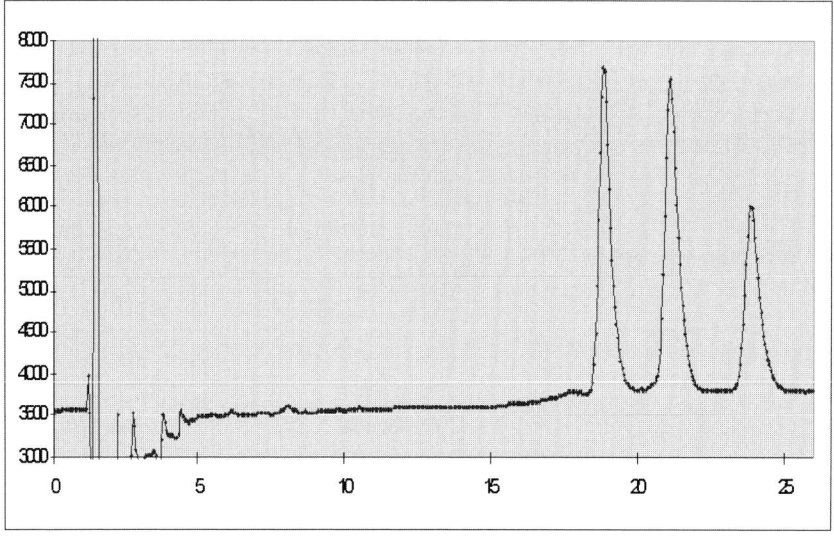

Figure 2-6. A chromatogram with 3 peaks [x=time(min), y=signal (µV)].

Table 2-5. Time windows from 12 to 12.95 min

Time (min)	Voltage (µV)
12.00	3603
12.05	3604
12.10	3603
12.15	3601
12.20	3602
12.25	3602
12.30	3605
12.35	3606
12.40	3605
12.45	3604
12.50	3602
12.55	3601
12.60	3600
12.65	3601
12.70	3602
12.75	3601
12.80	3603
12.85	3602
12.90	3603
12.95	3601

Table 2-6. Time window from 18.5 to 19.45 min

Time (min)	Voltage (µV)
18.50	3907
18.55	4101
18.60	4472
18.65	5064
18.70	5843
18.75	6664
18.80	7333
18.85	7666
18.90	7621
18.95	7269
19.00	6756
19.05	6220
19.10	5748
19.15	5361
19.20	5055
19.25	4810
19.30	4606
19.35	4433
19.40	4288
19.45	4168

The first derivation curve intersects the x-axis at the peak maximum. The intersection points with the second derivation are the points of inflection. In the time window of 1 min, for example, 20 slope values are used to give a mean value that is characteristic of the noise of the detector. The data system calculates the standard deviation of the voltage in the time between 12 and 12.95 min multiplied by 6, to get the three-fold standard deviation below and above the mean. The integrator operates similarly using the slope test. The probability that the slope is accidentally exceeded is less than 0.3%. If this were the case, then it would be the first indication that this could be the beginning of a peak. The start of the peak is only confirmed when the slope value (usually multiplied by a small number for safety) has been exceeded more than twice. The data system then defines the data point before the slope was exceeded for the first time as the beginning of the peak. This means that the data system has to go back a defined number of data points.

One should bear in mind that the setting of the sampling time (ST) – the time interval between two data points – is very important.

2.3.2.3 Setting parameters and their effect on peak area and peak height

In the above example of a "slow HPLC" from a 25 cm column, the ST is 0.05 min or 3 s. This parameter is more descriptive than the sampling rate (SR), which is the reciprocal of the ST:

$$ST = 1/SR$$

In our example, SR is 0.33 Hz.

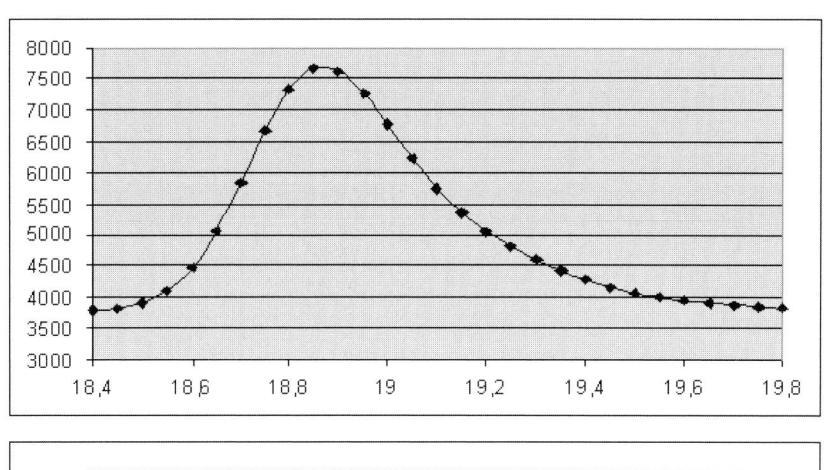
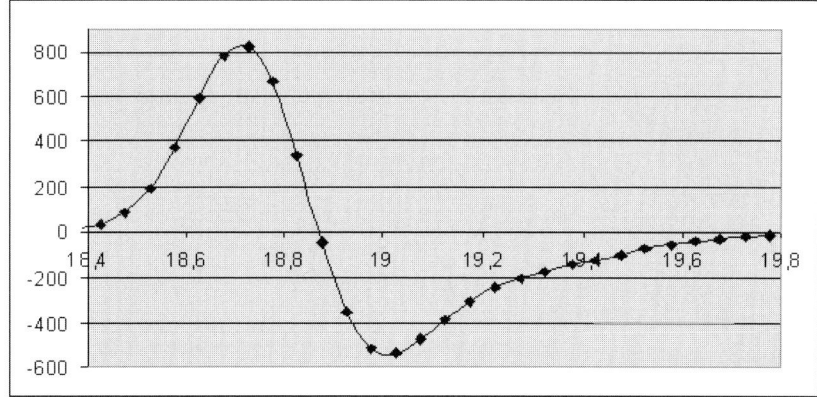
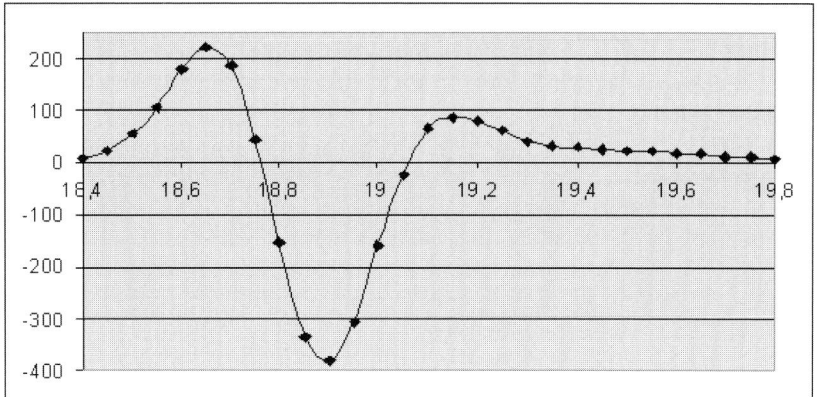

Figure 2-7. One peak, its first derivation as the difference of the data points and its second derivative as difference of the points of the first derivation.

Some integrators always scan the analogue signal with a high frequency that is also fast enough for fast capillary gas chromatography and which averages a defined number of data points before storing ("bunching"). This equalises the fast oscillations of the signal.

Let us assume that our data system collects the data at 100 Hz, namely 100 data points per second. This would require a bunching rate (BR) of 300 for a sampling time of 3 s. If the bunching rate is too high, the number of data points used to describe the peak is too low.

$$ST = BR/SR$$

The integration algorithm is only able to work optimally with the "correct" sampling rate. This value is often defined by the width, or more exactly, the width of the peak at half-height ($PWHH$). It is sufficient to describe the peak above the half-height with 10 data points. Thus:

$$PWHH = 10 \times ST \qquad ST = PWHH/10$$

The beginning and the end of the peak are described by another 10 data points so that the entire peak is described by approximately 30 data points. Some integrators need the width, some the sampling time or sampling rate. It is important to know the width at half-height of the smallest peak in the chromatogram in order to give the data system the "correct" sampling rate.

In some detectors it is possible to set a time constant to dampen the signal and therefore eliminate fluctuations. This time constant should never be higher than the ST. If an integration system is used, the time constant should be set to its lowest value to avoid falsifications.

Some integration systems make use of the second derivative to identify shoulders on peaks. A shoulder has no second maximum – obvious in the first derivative – but it does have two further inflection points.

The details of the very complex integration systems are given in the manuals. Ultimately, one has to believe that everything has been taken into consideration to obtain a good baseline in all cases. The user is primarily interested in the capability of the system, which should be tested for difficult separations and quantifications under realistic conditions, as described above. Integration systems are becoming increasingly powerful and more complex. This should not disguise the fact that the central function is the integration of peaks.

2.3.2.4 Where can mistakes be made?

The **time constant** must be no greater than one twentieth of the peak width, i.e., approximately a tenth of the peak width at half-height, and thus equal to the maximum ST. If the time constant is too low, it has no effect. While if it is too high, it may slur the peak and shift the retention time to higher values: the area does not change, but the peak height is smaller and the retention time longer.

The **width** should be approximately the peak width at half-height of the smallest peak in the chromatogram. This can be measured graphically or estimated from the area-to-height ratio. It is disadvantageous to make this value too high because the peak would be described by too few data points and this leads to unnecessary poor precision. It is better to use half the estimated value rather than the double. If the values are completely wrong, the integration algorithm cannot give good results.

The **sampling time** is one tenth of the width. The sampling rate is the reciprocal of the sampling time.

The **slope (threshold)** is usually measured in a region of the chromatogram that only shows the baseline. If the value is too small, a lot of peaks are recognised due to baseline perturbance, and this results in an endless report. If you can see that small peaks are not identified, the value may be too high. It is advisable to inspect the baseline in each chromatogram and to check it for plausibility. If in doubt, amplify or cut out with the mouse to inspect the baseline at the start and end of the peak.

One possible source of error in chromatography is the flow rate of the mobile phase. The effect of a constant flow is different for a mass flow-dependent detector (frequently found in gas chromatography), than for a concentration-dependent detector (HPLC detector without mass spectrometry). The mass spectrometer is mass-flow dependent.

Let us assume that we stop the flow for both types of detectors. In the concentration-dependent detector, the signal remains constant while the clock is running. Therefore, a large area is measured. In the mass-flow detector, the signal drops to zero until the flow starts again. If the two parts of the peak are reunited, there is no change in the area. This demonstrates the high dependency of concentration-sensitive detectors on the quality and constancy of the HPLC pumps. Asshauer and Ullner have shown large alterations of the area and small alterations of the height caused by changes of pressure and flow [5].

2.3.3 Methods of Quantification

2.3.3.1 What is the 100% method?

In a standard report, the area% or height% are normally issued. The areas (heights) of all peaks are added and then are assigned as being 100%. The area% (height%) is then expressed as a percentage of the peak area (height) relative to the sum. If the absolute amounts are not important, the %-values can be used directly. However, one should guard against the possibility of an additional peak in one of the chromatograms in an analysis series. This would alter the area% (height%) because the sum of the areas (heights) has changed.

2.3.3.2 What is the external standard method?

In a calculation using the external standard method, 5–10 samples with different (known) concentrations are measured. A calibration curve is obtained by linear regression. The equation of the straight line is transformed with known coefficients to allow the calculation of the concentration of the analysis sample from the measured signal to be made: $x=(y-a)/b$. If the intercept does not differ significantly from zero, you can also use a straight line through zero. Calibration with just one concentration can only be used for a straight line through zero. Some integration systems use a response factor (RF), which is equivalent to the slope of the straight line.

There are two useful (German) internet addresses on calibration that contain Excel spreadsheet masks for DIN 32645.

1. *www.ces.ka.bw.schule.de/lehrer/culm/praktikum/auswertung/kalib/kalib1.htm*
2. *www.rzuser.uni-heidelberg.de/~df6/tox/dintest.htm*

Enter the concentrations into A2 to A11 and name them "x". Enter the signal values into B2 to B11 and name them "y". Then enter:

|=INTERCEPT(y,x)| into B12

|=SLOPE(y,x)| into B13.

If these values are to be used in further equations, it is advisable to give them a name, for example, the text on the left in column A. Then you can enter |=b*x+a| into C2. Expand the calculation instruction to C11 to obtain the estimated signal values y_x.

Mark A2 to C11 and use the diagram assistant to create an xy diagram and the graph of a straight line.

	A	B	C	D
1	x	y	y_x	IS
2	2	5484	5212	20270
3	4	10280	10071	20366
4	6	15249	14929	20213
5	8	19503	19787	20099
6	10	23696	24645	19909
7	12	29546	29503	20573
8	14	34305	34361	20266
9	16	39687	39220	20427
10	18	43229	44078	19905
11	20	49764	48936	20080
12	a	354		
13	b	2429		

This straight line is estimated for calibration purposes. It gives the least-squared deviations between measured y and estimated y_x. It is also known as the squared residual standard deviation RSD:

$$RSD^2 = \Sigma (y - y\_x)^2 = \text{minimum}$$

$y = a + bx$

y = signal   x = concentration   a = intercept
b = slope
Mx = average of x
My = average of y

Figure 2-8. How to use Excel to calculate a linear regression.

The second address contains additional test data, estimated with different programs. These can be used to check the evaluation with an unweighted linear regression.

(The Figures 2-4, 2-8 to 2-14 including calculations are set up together and belong to one analysis series.)

2.3.3.3 Why use an internal standard?

An internal standard can be used to improve the quality of the analysis results. For example, if you inject the same sample manually into a gas chromatograph, you can't avoid variations in the injected volume. The addition of a defined amount of an internal standard to each sample will result, with high probability, in a decrease in the cv.

The idea behind this is as follows: if I find 1% more of the internal standard, I can assume that I have injected 1% more of the sample. This means that I now have to reduce the result by a factor of 100/101. This is probably not necessary if a highly precise autoinjector is used. The use of an internal standard is increasingly advantageous as the number of steps (often volume-transfer steps) in the sample preparation increases.

The internal standard must be added as early as possible so that all steps can be controlled. If a mass spectrometer is used as the detector, you can use the multiply deuterated analysis substance itself as an almost ideal internal standard. With other unspecific detectors, you should use a substance with very similar chemical properties for internal standardisation so that the extraction conditions or perhaps derivatisation conditions can be monitored as accurately as possible. A correction with an internal standard is only possible if the chemical behaviour is similar.

Another prerequisite is that the internal standard must give a baseline-separated peak. This is only possible if there is a spare gap in the chromatogram. It is possible that an internal standard cannot be used because it does not fit into the chromatogram, hence we are not able to solve the separation problem. The internal standard should be added in a small, but reproducible volume. The amount of internal standard should be in the region of the upper part of the standard curve of the substance being analysed.

The advantage provided by internal standardisation has to be proved. For example, compare the cv calculated for an external standard with the cv of an internal standard. If you can't prove a clear advantage, add it to your samples, but don't use it in the calculation. The use of an internal standard as a reference substance is worthwhile as you are free to use the internal standard if you change your mind. Calculating the cv of the internal standard itself can prove whether the internal standard adds more variance than it equalizes.

If you do decide to use the internal standard method, then do not use the areas (or heights) themselves for the calibration, but rather the areas divided by the areas of the

> The measurement of the internal standard (IS) is added to the data for yl and yu in Figure 2. The two columns y/IS contain the calculation |=A2/B2| or |=D2/E2|. It is obvious that the correction for the internal standard for yl (yu) results in a decrease of the cv from 1.35% to 0.65% (from 1.12% to 0.52%). Therefore, in this example, the use of an internal standard must be recommended.
>
> The internal standard is also included in Figure 6. Estimating this example in the same way leads to the same improvement. Therefore, the use of an internal standard must be preferred.

Wert	yl	IS	yl/IS	yu	IS	yu/IS
1	5443	20307	0.2680	49848	20033	2.4883
2	5323	20170	0.2639	50196	20228	2.4815
3	5326	20175	0.2640	51641	20710	2.4935
4	5377	20034	0.2684	51099	20746	2.4631
5	5362	20056	0.2673	50883	20456	2.4875
6	5198	19550	0.2659	50183	20090	2.4979
7	5294	19900	0.2660	50611	20523	2.4661
8	5221	19780	0.2639	50401	20325	2.4798
9	5311	20010	0.2654	50201	20342	2.4679
10	5294	20028	0.2643	49927	20255	2.4650
mean	5315	20001	0.2657	50499	20370	2.4791
stdev	71.6	216.0	0.0017	564.3	239.5	0.01279
cv(%)	1.35	1.08	0.65	1.12	1.18	0.52

Figure 2-9. The internal standard as a correction factor.

individual internal standards. The same is required for the signals of the analysis samples. All other steps of the calculation remain unchanged.

2.3.3.4 In which cases should the additions method be used?

Always calibrate with matrix samples – the matrix may be very simple, but usually it is not. Sometimes you cannot prepare a blank sample, for example, endogeneous substances in blood. In this case, it is possible to add a defined amount of the same substance and then measure at least one sample with and one without this addition. It is not favourable if the added concentration and the (unknown) concentration of endogeneous substance are very different because the result is obtained by subtraction.

0	12793
2	17469
4	22383
6	26782
8	30931
10	36593
12	41407
14	46775
16	50394
18	56893
b=	2420.83
a=	12454.5

To a sample containing 5 ng, the amounts in the first column were added and the signals of the samples were measured. The two coefficients of the straight line were fitted by an unweighted linear regression and the diagram was created. The amount without addition was calculated to be 5.15 ng (represented as a negative concentration). This value is in good agreement with 5.12 ng, which was calculated from the signal without an addition using the calibration described in Figure 7. A comparison of the two slopes shows very similar values. The conclusion drawn from a comparison of the normal calibration in Figure 6 and the result from this addition experiment is that the addition method has no advantage.

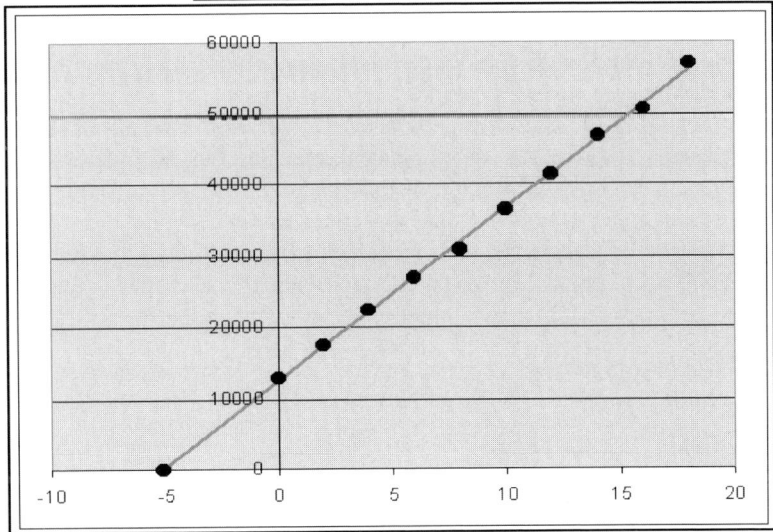

Figure 2-10. One example of the additions method in which no difference can be found compared with a normal calibration.

To a sample containing 15 ng, amounts between 2 to 18 ng were added. The least-squares calculation resulted in two parameters, a and b. The intercept, a, is the signal calculated for the sample without an addition. An amount of 15.1 ng was calculated by extrapolation to y=0. A comparison of the slope of the two examples for the addition method shows a large difference. Therefore, the addition method has a definite advantage in this case, comparing only the data from Figures 2-10 and 2-11.

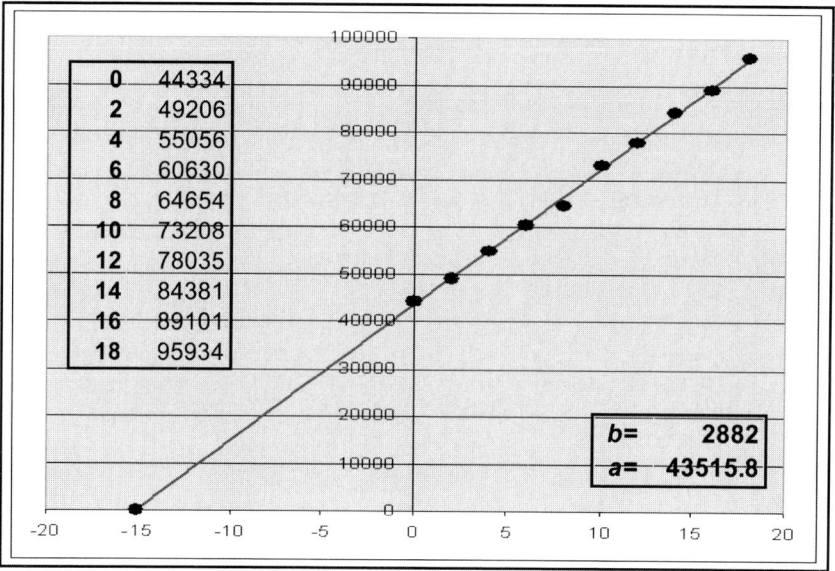

Figure 2-11. A second example of the additions method with large deviations from the first example.

Therefore, more than one concentration of added substance is commonly used – i.e., one individual calibration curve for each sample.

The calculated result using the addition method must be in the "working range" of the added concentrations. Otherwise the added concentrations are not adequate.

This clearly demonstrates the disadvantage of this method: you need a multiple of the number of analysis samples. If you find that different analysis samples give different slopes for the straight line, then the real advantage of the additions method become obvious: the matrix effects will be considered individually. This means that the additions method gives better results then normal calibration, if the matrix is variable. Some samples can only be analysed with the additions method.

2.3.4 Weighted Regression

2.3.4.1 What about the F-test? What are the other possibilities?

The *F*-test

The data of Figures 2-4 and 2-8 were measured in one series and belong together. Therefore, in addition to the data of the linear regression in Figure 2-8, we carried out a ten-fold measurement of the outer points of the working range. This gave the standard deviations $SDyu = 71.62$ and $SDyo = 564.31$. The *F*-value, calculated as the quotient of the squared standard deviations, is 62.08. Therefore, we have a very significant difference in the variances, that is, they are inhomogeneous.

What can we do?

1. Ignore the *F*-test.
2. Divide the working range into two similarly sized parts.
3. Use a weighted linear regression [6].

Case 1:
This is critical. It is common practice to use a (passed) *F*-test. The least-squares equations assume homogeneous variances.

Case 2:
Unfortunately, this is associated with a considerable amount of extra work. It may be necessary to split a very large working range of 100 or 1000 into more than two parts.

Case 3:
This method is not widely used, and is not required by DIN, EN or ISO standards. In spite of this, a weighting with $1/x$ or $1/x^2$ is often used to fit exponentials in pharmacokinetics. However, in simple cases, this is done by trial and error, and often intuitively with $1/x$ because $1/x^2$ seems too high.

Although weighting seems to have a coarse and arbitrary influence in the least-squares equations, this is not the case because the measurement of the two standard deviations shows that *SDo* is significantly larger than *SDu*. The normal linear regression is really weighting all deviations with a factor of 1. This means that the deviations at higher concentrations are taken into account too much, while those at low concentrations are not considered enough. Therefore, in the above example, the weighting cancels an incorrect supposition and is a return to reality!

In a concentration range of $(ABx=)$ 10, *F* has to be 10 (100) so that it is balanced by a $1/x$ ($1/x^2$) weighting. In our example, $F = 62.08$, that is, it lies between 10 and 100, but closer to 100. Therefore, a $1/x^2$ ($1/y^2$) weighting could be adequate. This corresponds to a weighting exponent $WE = 2$. It would be more precise to calculate the individual weighting exactly [7]:

$$WE = \frac{\log F}{\log ABy} = \frac{\log SDo^2 - SDu^2}{\log yo - \log yu}$$

In our example, the estimated weighting exponent would be 1.87. With this "model supposition", we have the additional advantage that the signal of an unknown analysis sample is associated with a defined weighting. Without a model this would be a problem. In our example, we balanced a nearly nine-fold difference in the standard deviations by multiplying SDu by 3 and dividing SDo by 3.

2.3.4.2 How do we weight the individual values?

It is better to base the weighting on the y values because the measured standard deviations, SDy, should be more closely related to y than to x. The weighting is performed in two steps. The preliminary weighting factor is given by:

$$g = \frac{1}{y^{WE}}$$

and g is normalised [8] so that $\Sigma w = n$:

$$w = \frac{g \cdot n}{\Sigma g}$$

Almeida [9] has pointed out that the equations for the normal linear regression can be converted into those for weighted linear regression "by adding a term w_i to any sum and changing any term n into Σw_i". Conversely, equations with weighting contain the normal (unweighted) case by changing Σw_i into n and setting all w_i to 1.

$$bw = \frac{\Sigma xyw - n \cdot Mxw \cdot Myw}{\Sigma x^2 w - n \cdot Mxw^2} \qquad aw = Myw - b \cdot Mxw$$

If you know the weighting factor, w, it is no problem to calculate aw and bw with only a pocket calculator. It is even simpler with Excel, as is demonstrated in Figure 2-12.

2.3.4.3 How to use Excel for weighted regression

Of course, it is not possible to use the Excel functions slope and intercept. Also only very few statistics programs are able to take a weighting factor into consideration.

In Excel you can arrange your calculation to suit. Figure 2-12 shows how to input the necessary calculations. This results in a spreadsheet (Figure 2-13), in which the estimated weighting factor $WE = 1.87$ was used. The corresponding diagram (use the diagram assistant) is shown in Figure 2-14. To recognise the effect of the weighting, the prediction interval was also calculated [10] and represented graphically:

$$y_i = y - x_i \pm t \cdot RSD \sqrt{\frac{1}{w_i} + \frac{1}{\Sigma w} + \frac{(x_i - Mxw)^2}{\Sigma w \cdot (x - Mxw)^2}}$$

x_i are the single concentrations of the calibration curve

The effect of weighting on the position of the straight line is less than expected. Without weighting, there is a constant band positioned around and parallel to the straight line. There is a slight widening towards the ends.

The weighting has the effect of lowering the "mobility" of the best fit of the straight line at low concentrations and increasing it at high concentrations, in agreement with reality. It is no wonder that the parameters of the straight line are barely influenced by the weighting.

> It is best to reserve the upper rows for the results so that the spreadsheet can be extended downwards. Let's write an identifier for the columns into row 6. Thus, type the following text cell-by-cell into A6 to I6: x, y, g, w, xw, yw, xyw, xxw, y_x. If, for example, a maximum of 24 concentrations is expected, mark A7 to A30 and enter the cell name into the name box. This name and not the text in A6 can be used in the calculation instructions (which always begin with "=") to simplify them and make them more concise. To see this name, click on the cell or mark the cell region. The text in row 6 is always visible and is used for visual identification. If you want to give regions of the spreadsheet a background colour, use <format> <cells> <patterns>.
>
> Type |=number(x)| into H1 and give cell H1 the name "n". In G1 you can enter (optionally) |n=|, which is only an explanatory text. Then into A1 enter |Regression with weighting exponent WE=|. Into cell E1 type "0" (zero) and name it "WE". Type |y=| into A2 and |x+| into C2.
>
> Row 5 is reserved for the sums of the variables between rows 7 to 30. Therefore, enter |=SUM(x)| into A5 and so on up to |=SUM(y_x)| in I5. Each cell should be named with "S" followed by the identification text in row 6, for example, Sx to Sy_x. Now enter |=If(y>0,y^-WE,"")| into C7 and expand this instruction to C30. You can then see the preliminary weighting, g, but only if a concentration is given in column A. Otherwise the cell is empty: (""). In D7 enter |=If(y<0,n*g/Sg,"")| and expand to D30. Column D now contains the final weighting, w, whose sum, Sw, must be equal to n.
>
> Enter |=If(y>0,x*w,"")| into E7, |=If(y>0,y*w,"")| into F7, |=If(y<0,x*y*w,"")| into G7 and |=If(y>0,x*x*w,"")| into H7 and expand to row 30. Now we only need the average of xw and yw to be able to calculate the intercept aw and the slope bw. Therefore, type |=AVERAGE(xw)| into E4 and name it "Mxw". Type |=AVERAGE(yw)| into F4 and name it "Myw".
>
> Enter |=(Sxyw-n*Mxw*Myw)/(Sxxw-n*Mxw*Mxw)| into B2 and name it "b". Type |=Myw-b*Mxw| into D2 and name it "a". Now you only need to enter |=a+b*x| into I7 and expand to I30 to obtain the spreadsheet shown in Figure 11.

Figure 2-12. Weighted regression with Excel.

The measurement uncertainty (mu) can be read from a graph of the prediction interval (pi) in the x direction [$pi(95\%)$ for 30 000: 11.8 to 12.6]. As the measurement of the standard deviation at the endpoints of the working range demonstrates, the uncertainty reflects the real world. In practice, every analysis result is calculated directly [11], including the mu:

$$x_a = \left(\frac{y_a - a_w}{b_w}\right) \pm \frac{t \cdot RSD}{b} \cdot \sqrt{\frac{1}{w_a} + \frac{1}{\sum w} + \frac{(y_a - Myw)^2}{b_w^2 \sum(x - Mxw)^2}}$$

y_a is the measured signal of the analysis sample, w_a the related weighting

Without weighting, the mu is too high at low concentrations [12]. In addition, the limit of detection and the limit of quantification are (unnecessarily) estimated too high, as shown by Castillo [10]. With weighting, the limit of detection is found to be very similar to the signal-to-noise method, which is realistic in chromatography.

The possibility given in some integration programs to weight with $1/y$, $1/y^2$ or maybe not at all is a decision that has to be made for the validation. It then applies to the lifetime of the method. The calculation of the weighting exponent shown here takes account of it as an individual variable for the description of the straight line according to slope and intercept. Therefore, it provides a greater flexibility of the analyb = =(Sxyw-n*Mxw*Myw)/(Sxxw-n*Mxw*Mxw)

	A	B	C	D	E	F	G	H	I
1	Regression with weighting exponent WE=				1.87		n=	10	
2	y=	2393.57243	x+	680.851704					
3									
4					4.32802803	11040.3003			
5	110	270741.5	1.7838E-07	10	43.2802803	110403.003	833606.156	335.957539	270101.484
6	x	y	g	w	xw	yw	xyw	xxw	y_x
7	2	5484.3	1.0182E-07	5.7081538	11.4163076	31305.2279	62610.4557	22.8326152	5467.99657
8	4	10279.9	3.1447E-08	1.76292198	7.05168791	18122.6616	72490.6465	28.2067516	10255.1414
9	6	15248.9	1.5043E-08	0.84332755	5.05996533	12859.8175	77158.9053	30.359792	15042.2863
10	8	19503.1	9.4952E-09	0.53230249	4.2584199	10381.5486	83052.3891	34.0673592	19829.4312
11	10	23696.4	6.597E-09	0.36982501	3.69825012	8763.52142	87635.2142	36.9825012	24616.576
12	12	29545.5	4.367E-09	0.24481245	2.93774942	7233.10629	86797.2754	35.252993	29403.7209
13	14	34304.5	3.3029E-09	0.1851594	2.59223164	6351.80073	88925.2102	36.2912429	34190.8657
14	16	39686.6	2.515E-09	0.14098998	2.2558396	5595.41275	89526.6041	36.0934337	38978.0106
15	18	43228.8	2.1434E-09	0.12015905	2.16286282	5194.33134	93497.9642	38.9315307	43765.1555
16	20	49763.5	1.6473E-09	0.0923483	1.84696598	4595.57457	91911.4914	36.9393195	48552.3003

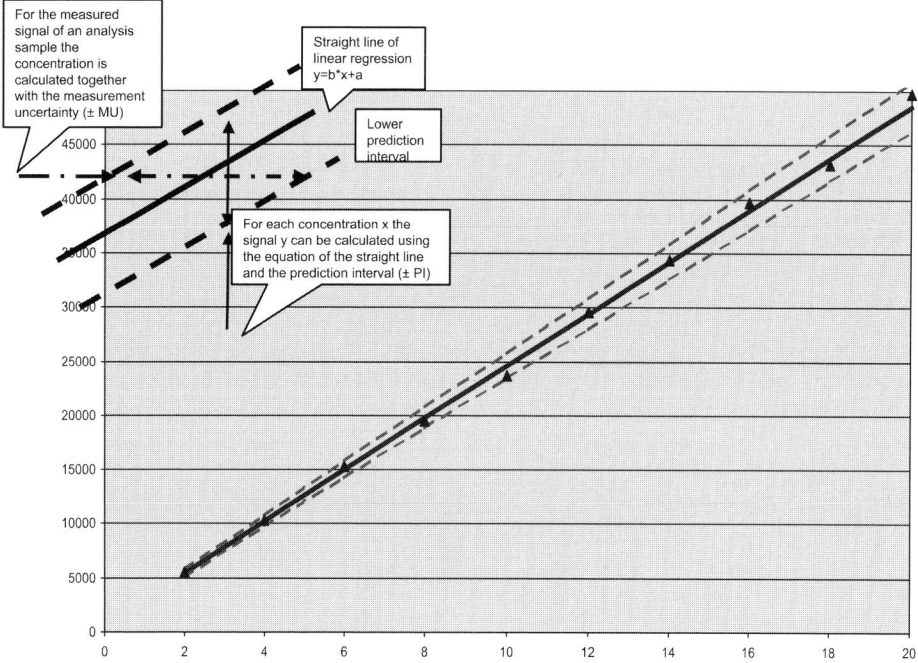

Figure 2-13. The example from Figure 6 with weighting.

Figure 2-14. Weighting in chromatography often results in a more realistic prediction interval and a more realistic uncertainty of the result.

sis method and better control. In each (back) calibration, the weighting exponent is checked. If the analysis results are used for pharmacokinetic purposes, one could think about the possibility of employing the weighting exponent to fit the concentration/time curve.

A weighted regression does not require too much additional effort and is more closely based on reality than a ordinary regression, which is included as a special case. It is to be expected that a weighted regression will be used more frequently.

References

1. Funk, W., Dammann, V., Donnevert, G. *Qualitätssicherung in der Analytischen Chemie*, VCH-Verlag, Weinheim, **1992**, p. 14.
2. Neue, U.D., Phillips, D.J., Walter, T.H., Capparella, M., Alden, B., Fisk, R.P. *LC-GC INT.* **1995**, *8*, 26–33.
3. Meyer, V.R. *Fallstricke und Fehlerquellen der HPLC in Bildern*, Hüthig Verlag, Heidelberg, **1996**, p. 94 onwards.
4. Dyson, N. *Chromatographic Integration Methods*, The Royal Society of Chemistry, Cambridge, **1990**, p. 141.
5. Asshauer, J., Ullner, H. in: *Practice of High Performance Liquid Chromatography*, Engelhardt, H. (ed.), Springer-Verlag, Berlin, **1986**, p. 83.
6. Miller, J.N., Miller, J.C. *Statistics and Chemometrics for Analytical Chemistry*, Prentice Hall, New York, **2000**.
7. Kuss, H.J. LCGC Europe 16(12), December 2003, 819–823.
8. Burke, S. *Sci. Data Managem.* **1998**, *2*, 32–40.
9. Almeida, A.M., Castel-Branco, M.M., Falcao, F.C. *J. Chromatogr. B* **2002**, *774*, 215–222.
10. Castillo, M.A., Castells, R.C. *J. Chromatogr. A* **2001**, *921*, 121–133.
11. Caulcutt, R., Boddy, R. *Statistics for Analytical Chemists*, Chapman and Hall, London, **1983**, p. 106.
12. Doerffel, K., *Statistik in der analytischen Chemie*, Deutscher Verlag für Grundstoffindustrie, **1990**.

2.3.5 Solutions to the examples of 2.3.1.4

Preliminary remark

We all have our own approaches to figures and mathematical operations. This is why I will describe 2–3 alternative methods, so you can pick and choose. You might find small discrepancies in the results due to rounding errors.

Example 1

Concentration

$$c_x = \frac{c_{st}}{A_{st}} \cdot A_x = R_{fc} \cdot A_x = \frac{12}{6000} \cdot 8000 = 16 \text{ mg L}^{-1}$$

Absolute mass injected

$$m = c \cdot V_{inj} = \frac{16 \text{ ng}}{1 \text{ μL}} \cdot 20 \text{ μL} = 320 \text{ ng}$$

Example 2

Required concentration – via the formula ...

$$c_x = c_{komp} \cdot \frac{A_{ist}}{A_{komp}} \cdot \frac{A_x}{A'_{ist}} = 8 \cdot \frac{4000}{8000} \cdot \frac{5200}{4200} = 4.95 \text{ mg L}^{-1}$$

... and with help of the "rule of three":

$$\frac{8000}{4000} = 2 \text{ and } \frac{5200}{4200} = 1.238$$

That means:

$$2 \,\hat{=}\, 8 \text{ mg L}^{-1}$$
$$1.238 \,\hat{=}\, c_x$$
$$c_x = \frac{8 \cdot 1.238}{2} = 4.95 \text{ mg L}^{-1}$$

Absolute mass injected

$$m = C \times V_{inj} = \frac{4.95 \text{ ng}}{1 \text{ μL}} = 10 \text{ μL} = 49.5 \text{ ng}$$

Example 3

1. External standard

The result can be obtained using the formula or simple ratios – in fact, the formula is nothing but a "rule of three" in disguise. You decide what you prefer.

A.
$$c_x = R_{fc} \times A_x$$
$$= \frac{200 + c_x}{213\,115} \times 127\,210$$
$$c_x = \frac{200 \cdot 127\,210 + 127\,210 \cdot c_x}{213\,115}$$
$$c_x = 119.382 + 0.5969 c_x$$
$$c_x = 296$$

B. Relate the area in question to the corresponding concentration, first to the unknown concentration c_x, then c_x plus the 200 mg added and work out c_x.

$$127\,210 \mathrel{\hat{=}} c_x$$
$$213\,115 \mathrel{\hat{=}} 200 + c_x$$
$$127\,210 \cdot 200 + 127\,210 \cdot c_x = 213\,115 \cdot c_x$$
$$c_x = 296$$

C. Take the difference between the two compound areas which result from the added 200 mg. Here, too, the area is related to the unknown concentration and then the difference is related to the added 200 mg in order to work out the unknown concentration c_x.

$$200 \rightarrow 85\,905 \quad (85\,905 \text{ results from } 213\,115 - 127\,210)$$
$$c_x \rightarrow 127\,210$$
$$c_x = \frac{200 \cdot 127\,210}{85\,905}$$
$$c_x = 296$$

2. Internal standard

Here, too, the result can be obtained either by using the formula or using ratios.

A.
$$c_{xi} = \frac{A_{x_i}}{\frac{A_{A_i}}{A_i} \cdot \frac{A_{st1}}{A_{st2}} - 1} = \frac{200}{\frac{213115}{127210} \cdot \frac{174832}{172703} - 1} = 287{,}4$$

B. Relate the correlation between sample area and internal standard area to the corresponding concentration, first unknown c_x and then c_x plus 200 mg added.

$$\frac{127\,210}{174\,832} = 0.727 \mathrel{\hat{=}} c_x$$
$$\frac{213\,115}{172\,703} = 1.234 \mathrel{\hat{=}} c_x + 200$$
$$\overline{1.234 c_x = 0.727 c_x + 200 \cdot 0.727}$$
$$0.507 c_x = 145.4$$
$$c_x = 286.8$$

C. If the internal standard yields a smaller area after adding 200 mg of analyte, the area of the component must be adjusted accordingly.

$$174\,832 \rightarrow 172\,703$$
$$127\,210 \rightarrow c_x$$

$$c_x = \frac{172\,703 \cdot 127\,210}{174\,832} = 125\,661$$

It has been established that the unknown concentration c_x yields an area of 125 661. After adding 200 mg, an area of 213 115 is obtained, from which you subtract the area of the unknown concentration (125 661). This gives you the area produced by the added 200 mg. You can now relate the concentrations to their areas and work out c_x.

$$(213\,115 - 125\,661 = 87\,454)$$
$$87\,454 \rightarrow 200$$
$$125\,661 \rightarrow c_x$$

$$c_x = \frac{200 \cdot 125\,661}{87\,454} = 287.4$$

Example 4

1. Injected mass of pure substance

$$m_x = C_x \cdot V_{inj} = \frac{1\,\text{ng}}{1\,\mu\text{L}} \cdot 50\,\mu\text{L} = 50\,\text{ng}$$

2. Yes, the internal standard makes up for the injection error. There are several ways of proving this. I will show just three possible ways.

A. The relative standard deviation of the areas (coefficient of variation) and of the area ratios [ratio of the area of the analyte (A_i) to area of internal standard (A_{ist})] are calculated. The relative standard deviation of the area ratios is smaller. Consequently the use of internal standards gives more precise results. Of course, in a real life situation, more injections (e.g., six) should be used. Using coefficients of variation to answer questions such as these is a method that can be statistically well documented and should thus be the method of choice.

$$V_k = \frac{\sqrt{\dfrac{\sum(A_i - \overline{A})^2}{2}}}{\overline{A}} = 2.9\%$$

$$V_k = \frac{\sqrt{\dfrac{\sum\left(\dfrac{A_i}{A_{st}} - \dfrac{\overline{A}}{A_{ist}}\right)^2}{2}}}{\dfrac{\overline{A}}{\overline{A}_{ist}}} = 0.65\%$$

B. As above, you begin with establishing the ratio of analyte and internal standard areas. Let us take component A as an example:

$$\frac{66\,123}{128\,345} = 0.515$$

$$\frac{68\,123}{131\,345} = 0.519$$

$$\frac{70\,123}{134\,345} = 0.522$$

Consider the mean values of the second injection – the absolute area (68 123) and the relative area (0.519). Define these values as 100%. Then we relate the deviations of the first and third measurements to these values, considering both the area and the area ratios. The percentage deviation of the ratio is less than that of the absolute area, e.g., 99.23% compared with 97.06%.

$$0.519 \triangleq 100\% \qquad 68\,123 \triangleq 100\%$$
$$0.515 \rightarrow 99.23\% \qquad 66\,123 \rightarrow 97.06\%$$
$$0.522 \rightarrow 100.58\% \qquad 70\,123 \rightarrow 102.94\%$$

C. Divide the mean values (of the areas and the area ratios) by the individual values. The smaller the deviation from unity, the more precise are the measurements. Here too we find the smaller deviation with the ratio, e.g., 1.008 compared with 1.030.

$$\frac{0.519}{0.515} \rightarrow 1.008 \qquad \frac{68\,123}{66\,123} \rightarrow 1.030$$

$$\frac{0.519}{0.522} \rightarrow 0.994 \qquad \frac{68\,123}{70\,123} \rightarrow 0.971$$

3. Extraction result

Define the area at the 100% extraction result (no sample preparation, no losses) and then at the x% result after extraction.

$$68\,123 \triangleq 100\%$$
$$49\,123 \triangleq x \Rightarrow A = 72\%$$

$$70\,234 \triangleq 100\%$$
$$51\,234 \triangleq x \Rightarrow N = 73\%$$

$$131\,345 \triangleq 100\%$$
$$98\,345 \triangleq x \Rightarrow P = 75\%$$

4. Working out the concentration of A and B in a urine sample

- External standard method for A:

$$c_A = R_{fc} \cdot A_A = \frac{50}{49\,123} \cdot 23\,123 = 23.5 \text{ ng}$$

For B:

$$c_B = R_{fc} \cdot A_B = \frac{50}{51\,234} \cdot 49\,234 = 48 \text{ ng}$$

For P:

$$c_P = R_{fc} \cdot A_P = \frac{50}{98\,345} \cdot 103\,345 = 52.5 \text{ ng}$$

- Internal standard method for A:

$$\left(\frac{49\,123}{98\,345} = 0.50\right) \quad \left(\frac{23\,123}{103\,345} = 0.22\right)$$

$$0.50 \rightarrow 50 \text{ ng} \quad\quad 0.22 \rightarrow c_A$$

$$c_A = \frac{50 \cdot 0.22}{0.50} = 22 \text{ ng}$$

For B:

$$\left(\frac{51\,234}{98\,345} = 0.52\right) \quad \left(\frac{49\,234}{103\,345} = 0.47\right)$$

$$0.52 \rightarrow 50 \text{ ng} \quad\quad 0.47 \rightarrow c_B$$

$$c_B = \frac{50 \cdot 0.47}{0.52} = 45.19 \text{ ng}$$

Comment:
P was found to be somewhat too high at 52.5 instead of 50 ng.

In a real case, this would have to be statistically proven. The measurements for A and B must thus also be too high, by the 5% that P deviates. The concentrations would have to be re-calculated to this lower level.

For A:

$$52.5 \rightarrow 50$$
$$23.5 \rightarrow c_{A \text{ corrected}}$$

$$c_{A \text{ corrected}} = \frac{50 \cdot 23.5}{52.5} = 22.4$$

For B:

$$52.5 \rightarrow 50$$
$$48 \rightarrow c_{B \text{ corrected}}$$

$$c_{B \text{ corrected}} = \frac{50 \cdot 48}{52.5} = 45.71$$

Another possibility would be to find a correction factor with which to multiply future results by – if a better internal standard cannot be found.

$$\frac{100}{105} = 0.952$$

For A:
$$0.952 \cdot 23.5 = 22.4$$
For B:
$$0.952 \cdot 48 = 45.69$$

We recommend checking the suitability of a substance as an internal standard early on in the process of developing a method. For this purpose, the internal standard is considered as a reference and its coefficient of variation is worked out. Thus, the quality of an internal standard can be assessed at an early stage.

Example 5

1. External standard method

Here, too, the areas are related to the corresponding unknown concentration c_x and $(c_x + 100)$ after the addition of 100 ng.

$$773\,241 \mathrel{\hat{=}} 100 + c_x$$
$$662\,328 \mathrel{\hat{=}} c_x$$
$$\overline{773\,241 \cdot c_x = 66\,232\,800 + 662\,328 \cdot c_x}$$
$$c_x = 597 \text{ ng}$$

2. Internal standard method

The areas are related to the corresponding amounts.

$$(3.86) = \frac{773\,241}{213\,277} \mathrel{\hat{=}} 100 + c_x$$
$$(3.11) = \frac{662\,328}{213\,277} \mathrel{\hat{=}} c_x$$
$$c_x = \frac{3.11 \cdot 100 + 3.11 \cdot c_x}{3.86}$$
$$3.86 \cdot c_x = 3.11 \cdot c_x + 311$$
$$0.75 \cdot c_x = 311$$
$$c_x = 415 \text{ ng}$$

$$c_x = (100 + c_x) \cdot \frac{200\,438}{773\,241} \cdot \frac{662\,328}{213\,277}$$
$$c_x = 100 \cdot 0.259 \cdot 3.105 + c_x \cdot 0.259 \cdot 3.105$$
$$c_x = 80.5 + c_x \cdot 0.804$$
$$c_x = 413 \text{ ng}$$

Another possibility would be to work with the differences as in Example 3, version C.

$$213\,277 \rightarrow 200\,438$$
$$662\,328 \rightarrow c_x$$
$$\overline{c_x = \frac{200\,438 \cdot 662\,328}{213\,277} = 622\,457}$$

and

$$(773\,241 - 622\,457 = 150\,784)$$

$$150\,784 \rightarrow 100$$
$$423\,765 \rightarrow c_x$$

$$c_x = \frac{100 \cdot 622\,457}{150\,784} = 413\,\text{ng}$$

Comment:

There is a great discrepancy in the results obtained by the external and the internal standard method, 597 ng against ca. 413–415 ng. In such a case, the measurement will have to be repeated several times in order to achieve a result of statistical relevance.

Case 1:

Repeat measurements yield a large variance, and the mean results obtained with the two methods differ greatly. This raises the following questions:

- Is the internal standard suitable?
- Are we perhaps below the limit of quantification (LOQ) and should one perhaps add twice the amount?
- Could there be a technical flaw?

Case 2:

The variance is okay, but the great discrepancy between the two findings (figures) remains.

There is probably an error in the systematic approach (e.g., a component of the matrix that has not been separated due to low selectivity is included in the integration of the relevant peak). This calls for an internal standard measurement.

Case 3:

There is great variance, and the great discrepancy between the two figures (or findings) remains. There have probably been several mistakes during the complicated sample preparation. In this case, too, the internal standard method would be preferable. Furthermore, the added amount was probably too small with respect to the endogenous amount, in our example 1:6 – a point that should be considered in standard additions experiments.

In GC with its volatile samples and in HPLC methods dealing with a complex matrix and complicated sample preparations (e.g., contaminated soil, faeces, fermentation broth, pre-column derivatization, extraction and precipitation) a great fluctuation of results can be expected. In such cases, one should always look towards quantification with the help of an internal standard.

3 Appendix

3.1 Solutions to the Problems

3.1.1 Crossword – the solution

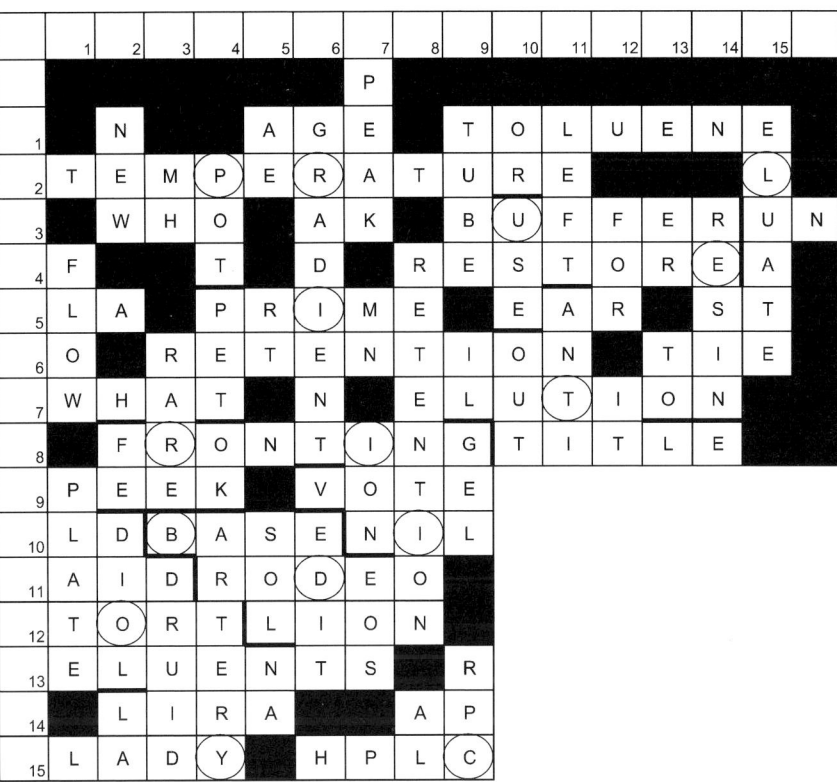

Reproducibility

3.1.2 An HPLC quiz – the solution

Table 3-1 gives the solution to the quiz, followed by brief comments.

Comments on 1:
When the quality of the packing declines it is possible that the peaks are deformed, i.e., the peaks may broaden and/or show tailing. Thus, the distance between the peaks at the base (i.e., the resolution) decreases. The quality of the packing has nothing to do with interaction (K) and selectivity (M).
Letters selected: L, E, I

Table 3-1

1. The packing has deteriorated		K	The peaks appear later
	X	L	The peaks become broader
	X	E	Resolution declines
		M	Selectivity declines
	X	I	Tailing appears
2. The proportion of acetonitrile in the eluent is increased		S	The peak area changes
	X	I	The retention time decreases
	X	C	The peak height changes
	X	E	The plate number changes
	X	Y	The lifetime of the column increases
3. The temperature is increased (ordinary RP-system)		B	Selectivity improves
		Z	Resolution improves
		W	The retention time increases
	X	T	Efficiency improves
		P	The retention factor is increased
4. The flow rate is increased		C	The peak area increases
		F	Resolution improves
		H	The plate number increases
		J	Efficiency improves
		O	Selectivity decreases
5. Endcapped C_{18} phases	X	S	... provide better peak symmetry for bases
		Q	... achieve a better separation of strong acids
		T	... achieve a better separation of bases, but they are unsuitable for non-polar substances
		X	... are more stable in an acidic eluent than non-endcapped C_{18} phases
		F	... mean that the surface is absolutely non-polar
6. A conditioning or saturation column (column between pump and injector)	X	V	... saturates the eluent with silica gel and protects the analytical column
		A	... must contain material with the same particle size as the separation column
	X	T	... raises the pressure
		K	... must be filled with the same stationary phase as the separation column
		C	... must also be thermostatically controlled in order to ensure the constant viscosity

Comments on 2:

When the elution strength of an eluent increases through an increase in its organic content, the peaks appear earlier (I), consequently they are higher (C) and narrower (E). The peak area is of course unaffected (S). As the polarity of the eluent decreases (higher ACN content), the polar silica gel is dissolved to a lesser extent, which prolongs the lifetime of the column (Y).
Letters selected: C, E, I, Y

Comments on 3:

When the temperature increases the peaks become narrower, and the efficiency increases (T). Selectivity usually deteriorates (B), which outweighs the advantage of the improved efficiency, so in the end the resolution, in general, will be poorer (Z). At the higher temperature, the peaks appear earlier (W). Thus, the retention factor (k value) as a measure of interaction decreases (P).
Letter selected: T

Comments on 4:

When the flow increases, the peak area decreases (flow×area=mass), which remains constant (C), the plate number – and with it efficiency – decreases (van Deemter equation!) (H, J). As a consequence, resolution also deteriorates (F). Selectivity is independent of the flow rate (O). It only depends on chemical parameters such as stationary phase, mobile phase and temperature.
Letters selected: none

Comments on 5:

Endcapping provides additional removal of some residual silanol groups [note the word some – it is impossible to make a surface absolutely non-polar (F)]. Bases are less retained, kinetics improve; so do peak shapes (S). Strong acids in their dissociated form do not hang around on a non-polar surface, so they elute early (Q). Non-polar substances do not have anything to do with the polar residual silanol groups (T). The short CH_3 groups used for endcapping are rapidly hydrolysed, i.e., acidic eluents make endcapped phases unstable (X).
Letter selected: S

Comments on 6:

V is the correct answer. After all, this is the purpose of such columns. Since this column is installed before the injector and thus not involved in the actual separation, particle size (A) and type of packing (K) hardly have any effect. The column oven (C) provides the constant temperature needed for consistent viscosity. There is usually no temperature control for the eluent before the injector – whether a saturation column is used or not. Finally, such a column provides some flow resistance, which increases the pressure (I).
Letters selected: V, T

Thus all of these letters give L,E,I,C,E,I,Y,T,S,V,T
Put in the right order, the solution is SELECTIVITY

3.1.3 An HPLC tale with Peaky and Chromy – the solution

- The choice of eluent was not the best – for the following reasons:
  1. For the separation of strongly polar components such as acids, a proportion of 85% ACN in the eluent is far too high.
  2. With 85% ACN and 100 mmol phosphate buffer there is a risk of salts precipitating in narrow capillaries. By the way, such strong buffers are not exactly conducive to a long lifetime of the column. An ionic strength of 20 mmol is usually sufficient unless you are working on ion-pair chromatography using older materials.
  3. If the pH value of 5 is correct (see below!) an acetate buffer would be more suitable, as phosphate is not really effective at pH 5. See Tip No. 27/1.
  4. Even more importantly, acidic eluents (about pH 2.5–3.5) must be used for the separation of acids. Thus, the acids would be un-dissociated and, as neutral molecules, be able to interact reasonably well with the C_{18} material.
  5. ACN is less suitable for the separation of acids – in spite of the narrow peaks, selectivity leaves much to be desired. Methanol is the better alternative in such cases, as the same elution strength yields better resolution.
- On a 125 mm×4 mm column and a flow rate of 1 mL min^{-1} the inert peak will appear after about 1 min. Here, however, the first relevant peaks appear far too early (low k values). There is no room for complacency as the conditions are anything but robust, and shifts in retention time are bound to happen.
- There is no need to use an endcapped phase for the separation of strong acids – it could even be a disadvantage. In an acidic medium, the short trimethylsilane group is fairly easily hydrolysed. The lifetime of endcapped phases in the acidic medium is also shorter than that of non-endcapped phases. Moreover, the selectivity of endcapped columns is in most cases insufficient for strong acids.
- Strong tailing has a detrimental effect on the detection limit and should not be accepted. Note that a tailing factor of 1.5 results in a decrease in peak height, which pushes the detection limit up by around 30%!
- Increasing the water content in order to prolong retention times is correct in principle, but when dealing with ionic substances it is far more effective to change the pH level or adding ion-pair reagents in order to modify retention time and selectivity.
- If only one of the compounds appears as a double peak, the cause cannot lie in the packing. If the packing is the problem it should affect all peaks. Here, however, it could be a case of prototropic equilibrium (a substance is present in its ionized and non-ionized form) or it could be a decomposition process. In any case, the problem is a component-specific one.
- A change in the proportion of water in the eluent can cause a change in peak height, but not in peak area.

3.1.4 Complete the sentences

Please continue the following sentences to turn them into valid statements. Sometimes there may be more than one answer.

In an RP system, ionic substances might elute very early. If that is the case I should:

- *– depending on the substance – use an acidic or an alkaline eluent for neutralization*
- *add ion-pair reagents to the eluent*
- *use a more polar RP phase, e.g., C_8, C_1, phenyl, nitrile or diol*

If only three out of eight peaks in my chromatogram are tailing:
I know that either the pH value of the eluent or the silanol concentration in the stationary phase has changed.

I can tell the packing quality of my column is okay because:
the theoretical plate number of neutral compounds has not decreased.

When I try and separate *bases* on Spherisorb ODS 1 or Hypersil ODS without using additives in the eluent I will get tailing peaks.

If in an eluent mixture methanol is replaced by acetonitrile at the same elution strength, the following will change in any case:

- *pressure (viscosity)*
- *lifetime of the column (lower pressure due to lower viscosity, lower dissolution of silica gel)*
- *peak shape (lower viscosity, thus faster kinetics)*
- *detection limit (lower viscosity, thus faster kinetics)*

and perhaps also the following:

- *retention time*
- *selectivity*
- *elution order*

Note that the greatest changes occur with ionic/polar substances.

Some operating protocols demand that selectivity between peak Nos. 4 and 5 be $a \geq 1.5$. This does not make sense. *It should read "Resolution $R \geq 1.5$ because the term resolution includes everything relevant for the separation, such as the strength of interaction (retention factor k), selectivity (separation factor a) as well as efficiency/separation performance (plate number N), see Tip No. 65. Then what use is good selectivity when the peaks are too broad, that means the number of plates is too low?*

Although the retention time has shifted I know that temperature and eluent are okay because:

- *the retention factor k has remained constant (a change in temperature and eluent would cause a change in the k value)*
- *the dead time t_M has also shifted (it is highly probable that a change in the flow rate has occurred)*

If I want to separate basic substances in an acidic medium I should expect them:
to elute very early, possibly with no retention at all.

The pressure has increased. On the basis of certain facts/pieces of information (left) I can exclude the following causes (right)

	To be excluded as causes for the increased pressure:
The dead time is constant:	*flow*
My column oven is working correctly:	*temperature*
The composition of the eluent has not changed:	*viscosity*
Eluent used methanol/water 50/50:	*salt precipitation*
I am working in recycling mode:	*the system has been clogged up by algae, fungi and bacteria*

3.1.5 "Matching pairs"

The area depends ...
... on the flow rate and the injection volume.

When there is a shift in dead time ...
... there is no need to prepare a fresh eluent.

The resolution depends ...
... on flow, stationary phase, temperature, particle size and the dead volume of the apparatus.

When you change your C_{18} supplier ...
... everything except the area can change (unless something is caught up).

When you change the composition of the eluent ...
... not only the retention time, but also the elution order may be changed.

Cold columns ...
... yield hot (good) separations.

Better a 5% error margin in the integration ...
... than 10% in the sample preparation.

When you raise the temperature ...
... selectivity (separation factor a) usually decreases while efficiency (theoretical plate number N) increases.

If you want sharp peaks ...
... you must have acetonitrile in the eluent.

Better sour cherries ...
... than an alkaline eluent.

Retention factor k ...
... does not depend on the flow or the length of the column.

Changing the pH value ...
... will change the retention time and peak shape of ionic substances dramatically.

Selectivity ...
... depends on temperature, eluent and the type of column packing.

Pepper in the soup is better ...
... than salt in the pump.

3.1.6 Did Peaky remember his lessons correctly – the solution?

In the following, Peaky's statements to amazed Chromy, and the correct answers with short comments where appropriate.

1. It makes sense to use endcapped phases to separate acids and bases. This will result in symmetrical peaks.
   False: Endcapped phases are only suitable for bases. In unbuffered systems, non-endcapped phases yield a better peak shape for acids. They are more stable at a more acidic pH value and for the most part are also more selective than endcapped phases.
2. Many of the latest C_{18} phases are not so suitable for the separation of strong acids.
   Correct: "Latest" stands here for well-covered hydrophobic phases.
3. Methanol/buffer eluents are better for the lifetime of a column than acetonitrile/buffer eluents of the same elution strength.
   False: Methanol is more polar than acetonitrile, which is why silica gel is more soluble in methanol.
4. Adding methanol or acetonitrile will cause the pH value of the eluent to drift into the alkaline.
   Correct
5. The greatest changes in selectivity usually happen around the pK_a values of the analytes, whereas the greatest robustness is achieved with a separation at a pH value that differs from the pK_a by ±2 pH units.
   Correct
6. When the pH value changes from 3 to 5, everything else remaining constant, the following parameters may change:
   False: The lifetime of the column is not affected while the other parameters may indeed change.
   – Peak height   – Lifetime of the column
   – Peak area   – Peak symmetry
   – Retention time   – Plate number
7. When permanently used in an acidic (pH ca. 2) or in an alkaline (pH ca. 10) environment, silica gel will slowly but surely dissolve.
   False: Silica gel is unstable in the alkaline. By contrast, it is stable under acidic conditions, even very acidic conditions. In the acidic pH range, short alkyl chains (C_8, C_3, C_1) may be hydrolysed or cleaved off, resulting in bleed from the column.
8. Selectivity permitting, one should work at a pH value of around 2.5–3.5 because many silanol groups are undissociated in this range. This reduces their interaction with polar compounds, and peaks become more symmetrical.
   Correct
9. KH_2PO_4 as buffer salt is less aggressive towards the column than $(NH_4)_2CO_3$.
   Correct

10. Ionic substances elute earlier when the ionic strength (buffer strength, salt concentration) is increased.
    Correct: Ionic (polar) substances remain longer in an eluent that has become more polar, because " like prefers like".
11. The pH value of the eluent should only be measured after the addition of methanol/acetonitrile, because a considerable drift could occur if they are subsequently added to the original aqueous phase. The final pH value may not even be known.
    False: Although the statement about pH value drift is absolutely correct, the conclusion is not. In solutions with an organic proportion of more than 40–50% measuring the exact pH value is almost impossible without specific electrodes. Such pH values have no practical value (e.g., in method transfer) unless they are just being used for comparative purposes. Most importantly, you want to know the pH relative to the pK_a of your buffer, in order to understand the buffer capacity of your mobile phase. See Tip No. 14.
12. Increasing the flow rate shortens analysis time but also increases the use of solvent.
    False: Volume is flow times time ($V = F \times t_R$) and remains constant, e.g., 1 mL min^{-1} × 10 min = 10 mL and 2 mL min^{-1} × 5 min = 10 mL. When the flow rate is increased and the elution volume remains constant, analysis time (t_R) decreases, and the separation can be finished earlier.

3.2 From Theory to Practice – Empirical Formulae, Rules of Thumb and Simple Correlations in Everyday HPLC

In this section I have compiled some rules of thumb, simple formulae, etc., taken from theoretical chromatography and hydrodynamics. Some of them can even be worked out mentally and thus provide a rough estimate as to the basis of which decisions can be made.
Examples:

- How high a pressure do I have to put up with if I want to reduce retention time by a factor of two by increasing the flow rate? (Example 7)
- What do I gain from a reduction in column length and particle size if I am happy with the resolution I am achieving with my system as it is? (Example 3)
- How much solvent do I save in real terms if I switch from my usual 4 mm to 3 mm columns? (Example 9)

Each formula or rule of thumb is accompanied by an example with figures. Where needed, a short comment is added. The first three rules of thumb were provided by John Dolan and Lloyd Snyder.

1. Dead time (breakthrough time, hold-up time, t_M), see also Tip No. 65
Empirical formula for estimating dead time if you cannot or are not allowed to add an inert compound to the sample, such as sodium nitrate, uracil or thiourea.

$$t_M \approx 0.1 \times \frac{L}{F} \text{ for 4.6 mm columns}$$

$$t_M \approx 0.08 \times \frac{L}{F} \text{ for 4.0 mm columns}$$

t_M = dead time in min
L = length column in cm
F = flow in mL min^{-1}

Generally applicable formula independent of inner diameter:

$$t_M \approx 0.5 \times \frac{L}{F} \times d^2$$

d = inner diameter of the column in mm

Example:
At a flow rate of 1 mL min^{-1} with a 125×4 mm column we expect the dead time to be around 1 min.

$$t_M \approx 0.08 \times \frac{12.5}{1} = 1 \text{ min}$$

and with the help of the generally applicable formula:

$$t_M \approx 0.5 \times \frac{12.5}{1} \times 0.4^2 = 1 \text{ min}$$

As columns are a bit more tightly packed these days, the real retention time of an inert compound is usually slightly shorter than the one estimated using this formula, but it still gives a good approximation.

2. Equilibration time in a gradient

$$t_{Eq} \approx t_M \times 0.15 \Delta\%B$$

t_M = dead time in min
$\Delta\%B$ = difference between initial and final percentage of organic component in the eluent

Example:
After a linear gradient of from 20 to 80% B (acetonitrile or methanol) the equilibration time for a 125×4 mm column and a flow rate of 1.5 mL min^{-1} can be estimated to be around 6 min.

$$t_M \approx 0.08 \times \frac{12.5}{1.5} \approx 0.67$$
$$t_{Eq} \approx 0.67 \times 0.15 \times 60 \approx 6 \text{ min}$$

This is a safe equilibration time when working with buffers or other additives in the eluent. If working with a straightforward methanol/water gradient, it may be even less.

3. Empirical formula for the dependency of the plate number on the length of the column and the size of the particles – see also Tip No. 60

$$N \approx 3000 \frac{L}{d_p}$$

L = column length in cm
d_p = particle size in µm

Let us look at a 150 mm column packed with 5 µm material and compare it with a 100 mm column packed with 3.5 µm material. Suppose both columns have been equally well packed, then the plate number should be roughly the same in both cases.

$$N \approx 3000 \frac{15}{5} \approx 9000 \text{ plates and} N \approx 3000 \frac{10}{3.5} \approx 9000 \text{ plates}$$

As retention time is proportional to the length of the column – see below – the shorter column provides the same separation (same packing material, same plate number), but saves one third of the time (150 vs. 100 mm), i.e., the analysis time is reduced by about 30%.

4. Resolution, selectivity (separation factor a), capacity (retention factor k), efficiency (plate number N) and retention time

This is the formula for resolution:

$$R = \frac{1}{4} \cdot \sqrt{N} \cdot (a - 1/a) \cdot \left(\frac{k}{k+1}\right)$$

N = theoretical plate number
a = separation factor
k = retention factor of the component that elutes later

The formula shows that for a good resolution the most sensitive function is selectivity: $R = f(a-1/a)$. This means that the first step in the optimization process should be an increase in a

Here are two examples with figures:

1. A successful increase of the separation factor a, e.g., from $a = 1.05$ to $a = 1.10$, the resolution is increased by a factor of 2!
2. For a baseline separation of two substances with a separation factor of 1.05, I would need a column with a theoretical plate number of about 6000. If I can successfully raise the separation factor to 1.10 – perhaps by changing the pH value – I would only need about 2000 theoretical plates for the same separation. I could now afford to increase the flow rate (faster separation, same amount of eluent used) or use a shorter column (faster separation, less eluent needed). Instead of a 100 mm column, for example, I could use a 30 mm.

$$N \approx 3000 \frac{L}{d_p}$$

Case 1 (separation factor $a = 1.05$): $L \approx \dfrac{N \cdot d_p}{3000} = \dfrac{6000 \cdot 5}{3000} = 100$ mm

Case 2 (separation factor $a = 1.10$): $L \approx \dfrac{N \cdot d_p}{3000} = \dfrac{2000 \cdot 5}{3000} = 30$ mm

However, if the separation factor is already 1.2 (which is quite respectable!) – and you are not happy with the resolution (because your peaks are apparently too broad), your next step should be to increase the plate number. See Tip No. 25. This is more economical.

Let us look at a different case!

The operation protocol gives you the information that the retention factor k (formerly capacity factor k') of analyte X is 4. The formula allows you to work out quickly how long it takes for the peak in question to elute.

Example:
You are using a 125×4.6 mm column and a flow rate of 1 mL min^{-1}.
The dead time can be estimated as follows:

$$t_M \approx 0.1 \frac{12.5}{1} = 1.25$$

$$k = \frac{t_R - t_M}{t_M} \text{ applies}$$

Solving the equation for t_R, yields:

$$t_R = (t_M + 1)k \text{ and thus } t_R = (1.25 + 1)4 = 9 \text{ min}$$

Incidentally, the k value is a multiple of the value of the dead time. How many times the dead time is the retention time? For example, for a substance that elutes after 4 min on a 125×4 mm column at a flow rate of 1 mL min^{-1}, the k value is 4. If at a flow rate of 2 mL min^{-1} the substance would also elute after 4 min, the k value would be 8.

t_M in the first case:

$$t_M \approx 0.08 \times \frac{12.5}{1} = 1.25 \text{ retention time equals 4 times the dead time: } 4 \times 1 = 4$$

t_M in the second case:

$$t_M \approx 0.08 \times \frac{12.5}{2} = 0.5 \text{ retention time equals 8 times the dead time: } 8 \times 0.5 = 4$$

5. Rules of thumb regarding the correlation between retention time and pH value, solvent and temperature

pH value

There is no general rule concerning the pH value. The shift in retention time depends on the pK_a value of the acid or base. In particular, with strongly ionic analytes, sudden changes in retention time by about 20–40% are fairly common when the pH value changes by 0.1 pH unit. If the analyte occurs in its undissociated form at a given pH value (i.e., it behaves like a neutral component) there will be – if any – only minimal changes.

Temperature

Rule of thumb: $\Delta 1\,°C \rightarrow$ 1–5% in the retention factor

Solvent

Rule of thumb: $\Delta 10\%\ B \rightarrow \Delta$ retention factor by 2–3

Example:

Suppose you are working with a 45/55 (v/v) acetonitrile/water eluent and find that the retention factor for a substance is $k=3$. A 47/53 mixture, which would differ by about 5%, would yield a k value of 4.5 (assumed change of k value by a factor of 1.5). This would mean that on a 125×4 mm column at a flow rate of 1 mL min^{-1} the retention time would increase from 6 to 9 min.

$$t_R = (t_M + 1)k$$

Case 1: $t_R = (1+1)\ 3 = 6$ min
Case 2: $t_R = (1+1)\ 4.5 = 9$ min

This is the reason why a difference of only 1–2% of the eluent composition often results in a shift of several minutes in retention time.

6. How high can the injection volume be without resulting in unacceptable peak broadening?

Assuming that the sample has been dissolved in the eluent or a solvent with similar elution strength, the following rule of thumb applies:

The injection volume should not be higher than 10% of the column volume. In this case, the contribution from the injection to the total band broadening is approximately 10%.

The volume of the column can be calculated approximately as follows:

$$V_k \approx \frac{t_M \cdot F}{0.8}$$

Or even more simply for 4 mm columns: $V_k \approx 0.1$ L

V_k = volume of column in mL
t_M = dead time in min
F = flow in mL min^{-1}
L = length of column in cm

Here are three examples with figures:
For a 125×4 mm column and a flow of 1 mL min^{-1}

$$V_k \approx \frac{1 \cdot 1}{0.8} \approx 1.25 \text{ mL or } 0.1 \times 12.5 = 1.25$$

For a 150×4 mm column and a flow rate of 1 mL min^{-1}

$$V_k \approx \frac{1.2 \cdot 1}{0.8} \approx 1.50 \text{ mL or } 0.1 \times 15.0 = 1.50$$

For a 250×4.6 mm column and a flow rate of 1 mL min^{-1}

$$V_k \approx \frac{2.5 \cdot 1}{0.8} \approx 3.13 \text{ mL}$$

Accordingly, you could inject 125 µL in the first case, around 150 µL in the second and around 300 µL in the third without significant peak broadening. Should the sample solvent be stronger than the eluent, see Tip No. 47. For critical separations, if you want the injection induced band broadening to be less than 1%, in such cases, I would recommend the following empirical formula:

$$V_{inj} \approx 0.2 \cdot \frac{t_R \cdot F}{\sqrt{N}}$$

V_{inj} = allowed injection volume
t_R = retention time in min
F = flow rate in mL min^{-1}
N = theoretical plates

For the first example in this section (125×4.6 mm column, flow rate 1 mL min^{-1}) and assuming a plate number of 9000 and a retention time of 5 min, the allowed injection volume is approximately 10 µL.

$$V_{inj} \approx 0.2 \cdot \frac{5 \cdot 1}{\sqrt{9000}} \approx 10.54 \text{ µL}$$

Hint: The higher the plate number, the lower the injection volume should normally be – in other words the smaller the particles or the better the column has been packed the less should be injected in order to make use of the high efficiency of the column.

We hold to the following:

The worse the column packing is and/or the later the peak is eluted then the injection volume is less critical. So, still using our example, and changing the retention time to 10 min, approximately 20 µL can be injected. See below.

If the packing quality decreases through long use, or when I have a badly packed column from the outset (e.g., 3000 theoretical plates), I can thus inject approximately 18 µL without problems – in both the last examples the peak will be somewhat broad ...

$$V_{inj} \approx 0.2 \cdot \frac{10 \cdot 1}{\sqrt{9000}} \approx 21.08 \text{ μL}$$

$$V_{inj} \approx 0.2 \cdot \frac{5 \cdot 1}{\sqrt{3000}} \approx 18.26 \text{ μL}$$

In conclusion, another empirical formula, to assist in directly estimating the allowed injection volume:

"The injection volume should be approximately 15% of the peak volume". Peak volume is the volume of the peak (to be more exact, for each measured component of the sample) found or resolved. The peak volume is simply the peak width at the peak base in minutes multiplied by the flow rate in mL min^{-1}.

So, with an allowed flow rate of 1 mL min^{-1} for a fast, relatively narrow peak with a peak width of 0.2 min, injection volume 30 μL and for a late eluting peak with a peak width of 1 min the injection volume of 150 μL should not in any case be exceeded.

$$V = F \times t = 1 \times 0.2 = 0.2 \text{ mL} \,\hat{=}\, 200 \text{ μL} \qquad \text{from which } 15\% = 30 \text{ μL}$$
$$V = F \times t = 1 \times 1 = 1 \text{ mL} \,\hat{=}\, 1000 \text{ μL} \qquad \text{from which } 15\% = 150 \text{ μL}$$

7. Correlations between retention time, pressure, length, flow, particle size and inner diameter of the column

The following relations apply:

$$t_R, P = f(L, F)$$

$$P = f\left(\frac{1}{ID}, \frac{1}{d_p}\right)^2$$

$$\frac{t_R}{w} = \text{constant}$$

$$t_R = f'(!)T, \, pH, \, \%B, \, I$$

In words this means:

Retention time and pressure are linearly proportional to length and flow.

Pressure has an inverse-quadratic relationship to particle size and inner diameter.

The ratio of retention time to peak width is constant for a particular compound under conditions that are not varied.

Finally, retention time depends on temperature and eluent composition, and for polar/ionic substances also on the pH value and the ionic strength.

The last relationships mentioned do not follow any unique mathematical function, but are generally exponential and also depend on the analytes and specific separation conditions. This is why it is difficult to make precise predictions as to what is going to happen in the optimizing process and why occasionally substances that differ in chemical character may elute in reverse order.

Three examples:

1. Increasing the length of a column from 125 to 250 mm results in a doubling of the retention time and pressure while the separation is improved by a factor of 1.4 (the

plate number is linearly proportional to the length, and resolution is proportional to the root of the plate number).

2. You are working with a 250 mm column, packed with 10 µm material, the flow rate is 1 mL min^{-1} and the pressure is 80 bar. In order to save time you want to use a 125 mm column filled with 5 µm material. What will the pressure be in the new column? It will be 160 bar. If you had only halved the particle size the pressure would have gone up by a factor of 4 to 320 bar, but since you also halved the length of the column, the pressure decreases again by a factor of 2. The result is 160 bar.

3. A 100×4 mm column, packed with 10 µm, produces a pressure of 20 bar. What would the pressure be in a column of the same length but an inner diameter of 2 mm, filled with 5 µm material? 320 bar, because the reduced inner diameter increases the pressure by a factor of 4 and because of the particle size, the pressure goes up by another factor of 4. Thus, the pressure is increased by a factor of 16.

8. Resolution, gradient duration and flow, see also Tip No. 63

A fictional case:

The development lab sends you a gradient method (linear gradient from 0 to 100% methanol) that has been developed on the hoof, so many things can still be changed. As usual, the specifications require a resolution $R \geq 1.5$, and the flow rate should be 2 mL min^{-1}. The gradient duration is given as 15 min. During the analytical process you notice that at this flow rate, the pump keeps stopping because its upper pressure limit is reached. Thus, the flow must be reduced to 1.5 mL min^{-1}.

- What gradient duration is required in order to achieve the specified resolution?

It should be 20 min because then the gradient volume would remain constant at 30 mL, and the resolution, in turn, depends on the gradient volume.

$$V_{grad} = F \times t = 2 \text{ mL min}^{-1} \times 15 \text{ min} = 30 \text{ mL}$$
$$\text{and } 1.5 \text{ mL} \times 20 \text{ min} = 30 \text{ mL}$$

- You notice that the last peaks elute even at 80% methanol. By how much can the gradient duration be reduced in order to save analysis time?

The gradient duration can of course be worked out using the formula in Tip No 63, but using a rule of three is easier still:

$$20 \text{ min} \rightarrow 100\%B$$
$$x \text{ min} \rightarrow 80\%B$$
$$x = \frac{20 \text{ min} \cdot 80\%B}{100\%B} = 16 \text{ min}$$

- The resolution achieved with a 250 mm column is far too good. It makes sense to change over to a 150 mm column. How long should the gradient run?

Again, the rule of three makes the answer easy:

$$20 \text{ min} \rightarrow 250 \text{ mm}$$
$$x \text{ min} \rightarrow 150$$
$$x = \frac{20 \cdot 150}{250} = 12 \text{ min}$$

9. Column diameter and eluent consumption

Table 3-2 states the savings in solvent that can be made from switching from a typical column diameter to a smaller diameter.

Note: It is practically impossible nowadays to find an HPLC instrument of such poor quality that it would not be possible to run it with a 3 mm column. Switching from a 4 mm to a 3 mm column would result in savings of nearly 50%. As long as the increase in pressure does not pose a problem it is well worth thinking about it.

Table 3-2. Savings in solvent

ID of the original column (mm)	ID of the new column (mm)	Approximate savings of solvent (%)
4.6	2	70
4.6	3	60
4	3	45

3.3 Information Resources for Analysis/HPLC

Preliminary note

The following selection of resources is by no means complete nor have any specific formal evaluation criteria been used. After all, every individual has his/her own way of using the Internet. It is up to the reader to evaluate the usefulness of the various sources.

HPLC books

New books on HPLC are being published all the time, covering general methodology as well as more specific areas. The relevant publishers and websites of special lab reviews give an overview. Here are some such websites:
www.wiley-vch.com
www.springer.de
www.amazon.com

Information on products and applications

The main manufacturers in HPLC not only offer product information on their websites, but also sometimes provide excellent general information and links to HPLC application databases. However, the applications shown there should always be viewed with the critical eye of the experienced user. The Internet has developed into an excellent source of even very specific information. In order to use it purposefully and efficiently, less experienced surfers are offered specific analysis-geared training courses – highly recommended!

Interesting analysis websites

The following websites plus short comments were compiled by two information service firms working in this area, whom I would like to thank for their efforts: Dr. Beyer Internet-Beratung and Chemie. DE Information Service GmbH.

1. HPLC-webtips by Dr. Torsten Beyer (www.dr-beyer.de)
   Overviews:

1. ACS LabGuide Online
   List of suppliers of lab products worldwide, including a large section about LC and LC/MS
   http://pubs.acs.org/labguide
2. Chemlin: (HPLC)
   Comments on HPLC links – mainly tutorials and literature
   http://www.chemlin.de/chemie/hp_lc.htm
3. Chirbase – A Molecular Database for Chiral Chromatography
   Database containing 85 000 chiral separations, growing at a rate of 4000 per month. Paid access only
   http://chirbase.u-3mrs.fr

4. HPLCweb

   Web portal featuring "Links", "News", "Education", "Literature", "Events", "HPLC Columns" and "Equipment"

   http://www.hplcweb.com

5. LC-GC Europe

   Homepage of "LC-GC Magazine" featuring specialist articles, reviews of new products and applications, troubleshooting, literature, useful links and a manufacturers' directory

   http://www.lcgceurope.com

6. Sciquest – Chromatography ColumnReSource

   Product database containing information on 90 000 columns manufactured by 90 manufacturers worldwide. Paid access only

   http://www.sciquest.com/columnshopper

7. separationsNOW

   Portal with suppliers' list, news, book tips, diary of events, forum, job offers, commented link lists and a newsletter dealing with HPLC, GC, electrophoresis and coupling technology

   http://www.separationsnow.com

8. www.chemie-datenbanken.de

   Overview of a range of general chemistry databases

Discussion forums

1. chrom-L – The chromatography email discussion list

   http://groups.yahoo.com/group/chrom-l

2. Discussing.Info

   http://www.discussing.info

3. Newsgruppe sci.chem.analytical

   http://groups.google.com/groups?oi=djq&as_ugroup=sci.chem.analytical

4. LC Resources Chromatography Discussion Group

   http://chromforum.com

2. Websites for analysts, compiled by Stephan Knecht (www.chemie.de)

Chromatography Forum [E]

The most extensive discussion forum on gas and liquid chromatography to be found on the Web. Moderators review forum contributions prior to publication in order to maintain the high quality standards of the forum. http://www.chromatography-forum.com (see above)

LC/GC-Magazine [E]

LC/GC-Magazine offers a good starting point for chromatography users, giving not only the latest news but also reviews of the latest products and software. A chromatography shopping guide helps find manufacturers and products.

http://www.lcgcmag.com

Chromatography Database [E]

Access to a wide range of HPLC chromatograms. The database can be searched by entering up to three chemical compounds. You can also focus your search by entering the type of detector, the mobile and stationary phase. A graph and a description of the analytical conditions are available.

http://www.quimifarma.net/crom/

Chemconnect – Journals and Magazines [E]
Over 600 hyperlinks to chemistry journals around the world, with alphabetical index.

http://www.chemconnect.com/library/journals/index.html or alternatively
http://www.chemconnect.com/about.html

... and finally:

A motely collection of websites relevant to analysis/HPLC I can recommend:

The first two websites can be searched entering up to three items in the first case, even more in the second.

www.interscience.wiley.com/cgi-bin/advancedsearche

www.link.springer.de/searche.htm

www.ingenta.com
Linking up various libraries and publishers worldwide

www.sciencedirect.com
Gives access to several reviews on chromatography

www.scirus.com
Science search engine run by Elsevier Publishers

www.logP.com
log$P$ values calculated according to substance structure

esc.syrres.com
Search for various substance data and other information using a range of criteria

www.chemistry.de
Look under databases to access a variety of databases in the field of chemistry

www.chromatographyonline.com
Articles from the journal "LC-GC" North America

www.chemistry.about.com
Informed links and also news on analysis

www.chemweb.com/analytical
Lots of interesting information on analysis

www.solventcentral.com
Physical and chemical data on solvents

www.saphirwerk.com
Cheap sapphire pistons for HPLC pumps

www.zirchrom.com
Information on buffers, dissociation constants, etc.

www.bi.umist.ac.uk/users/mjfrbn/Buffers/makebuf.asp
Automatic calculations for the preparation of buffers

www.mac-mod.com
Advice on choosing columns

www.forumsci.co.il/HPLC/Troubleshooting_Guide_Q_A.pdf
Good discussion of various points regarding troubleshooting. Reference kindly provided by Mrs Renate FitzRoy, St Andrews, UK.

www.acdlabs.com/columnselector
Possibility of finding similar columns

I would like to bring this compilation to a close on a reflective note.

1. Since we are living in an information society, it may be worth keeping the following aphorism in mind:
   "Acquiring information is cheap, disposing of it can cost you a fortune!"
   I think we should try and become more efficient.
2. Another statement seems also to capture the Zeitgeist or spirit of the times:
   "We are drowning in information without ever quenching our thirst for knowledge"
   The art is to turn information into knowledge by processing it. This is where the personal environment can have a catalytic effect, and using it efficiently means good communication. This will not only improve interpersonal relations in a company, but there are also very pragmatic reasons to put real communication at the top of your personal as well as your company's priority list, for your long-term as well as your short-term goals.

3.4 Analytical Chemistry Today

This section gives an – admittedly subjective – description of general current trends and their effects in analytical chemistry, underpinned by examples. Section 3.5 will focus on HPLC trends in particular. I do not intend to make a moral judgment – after all, we are all part of an ongoing process and may be perpetrators one day and victims the next. I am simply trying to establish facts and document my observations – hoping that lessons can be learned as far as our own responsibility is concerned.

The situation in the laboratory

We live in the age of globalization and turbo-capitalism. These terms may not be particularly beautiful, but they do describe the global economic situation accurately and objectively. Although there has been some resistance here and there, these are large-scale phenomena that have been shaping our patterns of thought, action and behaviour throughout, especially in business life. Let us restrict ourselves to the effects they have had on the chemical and pharmaceutical industry and on analytical labs in particular!

First problem – time

The focus on verifiable short-term economic success leads to ever-shorter product life cycles, with the inescapable result of ever increasing time pressure. This obstructs sensible, well thought-out courses of action.

Here are some examples:

- Well-established mature techniques – e.g., miniaturization – that could help cut costs and save time are not being used because there is no time to come to grips with them.
- There is no time – or to put it more bluntly, because we or the person in charge sets the wrong priorities – we do not take the time to think our projects through. It is a matter of ticking boxes rather than dealing with the matter at hand. Thus, a method is rashly adopted or a transfer plan "read through" – i.e., skimmed through. In both cases, one critical glance would have been enough to recognize that the method would not work in the real world of lab routine.
- We have become used to being in a constant rush and acting like headless chickens, which is easy to do in an environment that behaves in the same way. We have been tacitly accepting that the predominant objective "maximizing profits in the shortest possible time" does not only lead politicians astray. In other words, the benchmarks of decency and self-respect are being constantly lowered. Here are two examples, taken from everyday life:
  - A new column that can separate a wider range of degradation products due to its better selectivity is removed from the instrument and replaced by a column from the seventies. This last-century technology column does fulfil the formal requirements of providing a peak – albeit just one. Who has the time to fill out a document control form anyway?
  - The dead volume used in the formula is just a figure that will fit in nicely with the k value stipulated in the SOP.

Anybody who questions such procedures will at best be greeted with indifference, impatience or condescension. Often, even the neutral act of questioning is seen as

an act of insubordination. After a few frustrated attempts, most people give up. However, yesterday's best practice need not be right for today, and an attitude of "constructive discontent" would be the real key to success.
- Furthermore, time pressure leads to a dangerous emphasis on formal criteria and creates a mindset that focuses on superficial, easily obtainable evidence. Looking for quality assurance criteria rather than true quality is much easier under the current stressful conditions. It is always easier to prove a trivial point than to stand up for innovation in the face of tradition. What are the consequences?

The spirit of our time manifests itself in the collection of what is known as hard facts – signatures, figures, water-tight contracts, standard deviations and correlation coefficients. While focusing all our thoughts, time and energy on them, we hardly have any resources left for the "soft facts", such as strategic orientation, intuition, creativity, sensitivity and courage. These are the real keys to success, and to develop them, every person only has a limited supply of time, energy, etc.

Wealth is created not by being economical but by promoting good ideas.

What gives somebody the edge over a competitor is not a low standard deviation, but integrated interdisciplinary thinking, and these indispensable prerequisites of success are now in decline. They have become suspect or even the butt of a joke in the eyes of our modern busybodies. What was meant to be the – albeit important – framework to secure the future has become its objective, the be-all and end-all. Our efforts will go no further than our targets – the contract has been sorted out, the correlation coefficient is 0.999, the expert who will sort out things has been hired, the programme – whatever its content – has been agreed upon, a method has been submitted, etc.

Let us summarize: Life is much easier for those who can produce visible results, such as measurements, statistics or diagrams than for those setting aside time for reflection and improvement – things that cannot always be measured and – *horribile dictu* – may come to nothing. What a disaster in an environment where error is synonymous with weakness! Curing symptoms rather than the underlying disease is topical, cheap and does not pose any risks. Two examples will illustrate the point before we proceed to the next topic.

During an audit, it is easier to check whether the refrigerator has been adapted to specifications than to investigate how the variance of a method relates to the specification criteria.

In order to make an existing method faster, more robust and cheaper, if you are lucky, all you need to do is write up a document control report and perhaps revalidate, but if you are unlucky, you have to fight invisible internal and external enemies. Since you have neither the time nor the nerve to persist and your co-workers have enough samples to work through as it is, you will not bother. If you are interested in some proof of your success in order to be promoted, spending your time on increasing the number of samples or with laboratory co-workers is a far more successful strategy.

Second problem – time and short-term ROI (return on investment)

Lack of time, apathy and ROI-oriented short-termism must lead to a situation where networked thinking and action is replaced by individual isolationism and compartmentalized thinking.

Here are some examples:

- Since nobody has been prepared to take any risks, hundreds of millions or even billions of hours have been wasted in quality assurance labs over the past 20–25 years because a flow rate of 1 mL min^{-1} and/or 250 mm columns have been used in assay analysis. Why not be slightly more adventurous and set the flow rate at 1.5 mL min^{-1} or even – perish the thought – at 2 mL min^{-1}? I can understand that 1 mL min^{-1} is used in ancient methods, but even recently developed methods often stipulate the same old flow rate, and an analysis time of 12–15 min is still deemed acceptable for just 3–4 peaks.
- We may talk to one another, but we do not communicate efficiently, as the following example shows.
  You drive 400 km to take part in a seminar, and when you arrive you find to your surprise that two of your colleagues made the same journey in their individual cars. All seminar bookings were made by the same secretary. I have seen it happen more than once, and not only in big companies . . .

It is the exception rather than the rule that there is effective communication between the developer and the routine user of a method in the initial stages of a project. We all know the consequences. Let me give you the following example!

A development department has developed and validated a low-pressure gradient method. The control lab only possesses a new high-pressure gradient instrument. It goes without saying that the retention times differ, but since the method has already been validated, nothing can be changed.

It was therefore decided that the same low-pressure gradient instrument the development department had been using should be bought costing 55 000 €. Under the given circumstances, this seemed to be the cheapest and least risky policy. Although I understood the reasons behind the decision, I asked if it would not have been possible to contact the colleagues in the development department beforehand. The answer was just what I expected – a wry smile.

- The exact amount of an active ingredient has to be given right down to the second decimal point, and great care is taken that the coefficient of variation is not 1.6% but 1.5%, as required, but the leaflet that comes with the medication reads: "Adults should take 2 to 4 tablets daily, depending on their body weight". Well, analysis is one thing, marketing the product quite another . . .
- By switching the raw material supplier, the purchasing department has made a saving of 20 000 €. This was in line with the target savings set at 10% by the management board. While the purchasing department thus managed to reach its target, the loss caused by rejects went up by 200 000 €, due to insufficient purity of the raw material.
- In sensitive, GMP conforming workspace it is on the one hand quite clear that all the employees and visitors must pass through air locks. Lab coats will of course be disinfected etc. Also, possibly or definitely contaminated samples or other materials will only be handled with disposable gloves. On the other hand, employees from external cleaning companies are in practice rarely required to follow any shict disposal procedures for these areas. Contaminated waste is removed "as is", and waste bins and other containers are handled without precautions. And the untrained eye

cannot tell the difference between flour and any other white powders "But what have we got to do with this? And which of us is responsible?"

- A validation process is rushed through in two weeks using standards because the submission deadline cannot be changed. Any analyst will know that validating a method under real life conditions at such short notice is not serious science. The costs caused by reclamation, repeat measurements, putative or *de facto* SOS situations are spiralling up, weighing heavily on the controller's budget who will have a lot of explaining to do, etc. ... However, since this is not a book about cost analysis, I only want to point out that these are not production costs but avoidable error costs that do not appear on the balance sheet. What does not appear in the books either is a figure that accounts for frustration, loss of motivation, etc. of staff. Unused human resources are the dead capital of every firm and can amount to many times the value of storage, current assets, etc. which are the subject of so many board meetings and strategic planning sessions. This shows again just how flawed decisions taken in companies and elsewhere often are. It seems different when we let the people develop their opportunities – when not from conviction, at a minimum on the basis of ethical capitalism: only those who build upon self responsibility have long term success.

More examples could be easily added to the list, but I think I have made my point.

From our own experience, we all know how easy it is to get stuck in a rut, but the situation is currently exacerbated by a mindset that makes it difficult even to use well-established and recognized methods effectively. If the situation persists, company success could be in jeopardy. I am convinced that if a company could change its policy in any of the cases mentioned and adopt a longer-term perspective, considerable sums could be saved or gained, not to mention the improvement in staff morale. Let me just mention one point. A leader of a team or company unit who has established reasonably good long-term communication within his or her group will be rewarded in commercial as well as staff-motivational terms. I am not talking about high-handed decisions to send team members to conflict-handling and communication workshops, but of genuine communication between team leaders and colleagues as part of their daily life.

Pessimists (realists?) may find some consolation in the idea that the problems discussed affect us all, and people have always wanted to be seen to be doing something – however ineffective, as the following quotation suggests:

"We trained hard ... but it seemed that every time we were beginning to form up into teams we would be reorganised. I was to learn later in life that we tend to meet any new situation by reorganising; and a wonderful method it can be for creating the illusion of progress while producing confusion, inefficiency, and demoralisation".

This very topical comment was written by Petronius Arbiter around 65 A.D. It just demonstrates how little has changed over the last couple of thousand years. How much of a comfort this really is remains a matter of opinion.

3.5 Trends in HPLC

The hardware

Pumps, injectors and detectors

In a technology that has matured over the years, you do not expect quantum leaps to be made in developments with the equipment. What we see in the latest HPLC modules is young, fresh and compact designs and often some small but interesting improvements. Examples are the long UV cells with considerably increased sensitivity – by a factor of up to 5 – or intelligent injectors that only begin to inject once the preset injection volume has been verified. While the hardware has become more robust and thus more service-friendly, there is now far less the user can do to correct a problem unless a defective unit can be exchanged as a whole.

Columns

1. Column chemistry

"New RP Columns Suitable for Pharmaceutical Analysis" may sound like a quotation from a catalogue dating from the early eighties, the gold rush period in phase technology. Let me tell you that ever since the days of Tswett there has not been a single symposium on chromatography that did not see the introduction of a host of new materials. This may come as a surprise, but somehow, in spite of the huge development costs, it seems that there is still money to be made – or is there? We shall see. The trend of the nineties is continuing – and while still churning out the odd classical hydrophobic RP phases, most manufacturers seem to focus on very specific, mostly polar phases on the one hand, and phases that provide good selectivity for both polar and nonpolar analytes on the other. These hybrids are a compromise between two extremes – "RP" phases of a polar character that resemble silica gel and hydrophobic RP phases that resemble a purely polymer matrix. Let us take a closer look:

- Monoliths

Monoliths have finally been established as a valid alternative to the classic particles. They will become increasingly important in analytical LC as well as in capillary LC, but also in preparative or even process LC. As monoliths on silica gel bases have already been patented, firms are now developing organic monoliths, which – just like the classical columns – will not be of any risk to the position of silica gel. Hybrid material has also been widely accepted now, because separations in the alkaline remain an attractive subject.

- Embedded phases and others

New embedded phases are being developed that vary in polarity, depending on the embedded group, ranging from carbamate, urea and amide to ether, ethane and an ion-pairing group. The last renders classical ion-pair chromatography superfluous. Less prominent, but still widely discussed are hydrophilic endcapped phases. The objective of the manufacturers can be put in a nutshell: "We would like to provide you with the good selectivity of the polar materials of the seventies and early eighties using 'cleaner' silica gel, better batch-to-batch reproducibility and higher stability in routine analysis".

```
         HPLC

 Fast Analysis?  The "Three" ...

 ⇒  3 µm           Material
 ⇒  3 mm           Column (i.d)
 ⇒  3 cm           Column length
```

Figure 3-1. Column and particle dimensions for fast HPLC, explanation see text.

Zirconia and titania keep popping up in the discussion as alternatives to silica gel matrices for separations in strongly alkaline media or at higher temperature.

2. Column dimensions

As would be expected, the choice of available columns is huge. Figure 3-1 is based on a slide I produced in 1984. Even in those days, using a 3/3/3 column – 3 cm length, 3 mm inner diameter, 3 µm particles – did not pose a technical problem. However, there are still people around who, seeing this old slide, assume that it shows the technology of the future, while in other areas the use of 20–30 mm columns is long-established practice – see below. In these research environments, micro- and nano-LC will soon account for 30–50% of all separations. This upward trend is mostly due to proteomics, metabolomics and LC-MS coupling.

For most users claims such as "60 peaks in 100 ms" or "1000 mm×50 µm capillary LC-CE-MS-coupling" will – if ever – become relevant in 5–8 years at the earliest. However, the following dimensions seem to be practical in daily routines as long as the matrices do not cause problems and the instrument is handled with care: 20 mm length×2.1 to 4,6 mm inner diameter, 1.8–2 µm particles. These columns could be seen as some manufacturers' answer to patented monoliths when it comes to fast separations.

Technologies of the future

In years to come, miniaturization and coupling techniques will remain the prevailing topics – apart from more peripheral subjects such as sample preparation. Miniaturization will remain important because time saving and sensitivity gain will remain key objectives in the future. In coupling, one objective is to increase resolution (coupling of two methods, e.g., LC-GC or LC–CE), the other is increasing specificity by coupling HPLC with spectroscopy – e.g., LC-MS or LC-NMR. Or perhaps you want to improve both, in which case two separation methods might be combined with spectroscopy, e.g., LC-GC-MS, gel electrophoresis-LC-MS, IC-LC-MS or LC-GPC-FTIR.

Coupling

Looking at the various coupling methods, LC-MS(MS) is an already well established and mature process that is gaining significance, being routinely used in areas such as bioanalytics. It will probably make its way into more and more areas. LC-NMR, by contrast, is becoming established rather slowly. Dyed-in the-wool spectroscopists quite understandably want to take their measurements offline and prefer two-dimensional NMR to LC-NMR coupling which puts new constraints on both techniques. The "intelligent" splitting process according to which the substances can be directed to NMR, MS, and/or fluorescence detection is notoriously unreliable.

Examples of couplings that could become more important in the future:

- LC-NMR and LC-NMR/MS
- CE and CLC-NMR
- Two-dimensional nano-LC-MALDI-MS-micro fraction collector
- Multidimensional on-line sample preparation ("clean up") – LC-MS
- Multidimensional chromatography plus spectroscopy, e.g., strong cation exchanger > nano-LC (capillary with RP packing) > MS-MS

From today's perspective we can come to the following conclusions about current and future developments in HPLC:

- Silica gel as the matrix and RP as the mode will remain the number 1 in HPLC.
- In everyday lab practice, the most widely used particle size will be between 3 and 4 µm, while 2 µm material will remain an interesting niche product.
- Monoliths will become more widely available and cheaper, and they will offer a wider range of functionalities. The applications range from capillary LC and LC on a chip to preparative chromatography (e.g., 200 g protein; flow 2 L min^{-1}; column volume 8 L).
- Molecular imprints will increasingly replace the classical affinity media.
- Proteomics, metabolomics, diagnosis, etc., make CLC and nano-LC attractive options that will be developed further.
- LC-MS coupling will expand into more areas, and LC-NMR will be invaluable where new compounds or impurities are investigated.
- Multidimensional chromatography – perhaps with subsequent spectroscopy – is becoming increasingly important in sample preparation as well as in separation.
- Columns gain in stability and versatility while batch-to-batch reproducibility is improving, even for polar phases.

The software

Let us take a closer look at the software, specifically the part that is involved in integration, peak identification and data analysis. Apart from the column, this is what the user has to deal with most in a daily routine.

Originally an improvement on the humble electric typewriter, the PC has mutated into a jack-of-all-trades in the office. A similar development – albeit on a more modest scale – has taken place in chromatography data handling programs. More and more functions have been added to the classical process of chromatography (peak area and height, percentage area, asymmetry factors, plate numbers, etc.). Here is a list:

- Supervision and control of other equipment including GC
- All but complete validation
- Unrestricted functions for importing/exporting pre-treated or raw data, easy communication with a range of other programs
- QA-requirements met (GLP, CFR Part 11 etc.)
- Hierarchical structure of user levels, such as administrator, supervisor, user

Furthermore, the manufacturers are trying to make their products more attractive by introducing further add-ons.

Here are some examples:

- Taking into account data that give information about the state of the column, such as pressure, peak width, retention time, etc., plus individual user settings, a program may send reminders along the lines of "regenerate column" or "replace column".
- Simple optimization steps as you would find them in professional optimization programs are carried out by the software automatically.
- Troubleshooting programs contain video sequences that explain simple repair steps.
- Online repairs via a modem can be carried out by service engineers, or the user can get precise repair instructions via the internet.
- There are initial cautious attempts to predict the chromatographic behaviour of a compound based on available physicochemical data.
- Data analysis outside chromatography, such as principle component analysis of a large and complex data set, for example to find metabolites in urine matrices.

Obviously, the suitability of any software depends on individual requirements. In this sense, my comments made in Volume 1 are still valid, I regret to say. Some software packages may offer too much for your actual purposes, which makes them unwieldy and user-unfriendly. Far too little thought is given to the perspective of the analyst/chromatographer, and the special requirements of the two major application areas – routine analysis and method development – are barely taken into account when new HPLC software is conceived.

From my point of view, the ideal software consists of a basic package with two additional modules to choose from – a routine module and a method development module. These would have to be made compatible with each other in order to facilitate method transfer. A software manufacturer who could meet these needs would probably increase his sales considerably. When it comes to evaluating chromatography software you should make a personal checklist to see to what extent the program meets your needs before you make the purchasing decision. Here are some examples:

- Is it easy to change a method parameter quickly or do superfluous GLP requirements get in the way, making it necessary to open several windows to document/save/update changes in X/Y/Z?
- Is there enough memory left if everything has to be saved? Does the backup of raw data work without hitches or can data be lost?
- Are AD-converters/interfaces sufficiently protected against electronic interferences?
- How many different LC instruments (and what types of LC instruments) or – if required – GC equipment from other manufacturers can be run with this software without running into difficulty?
- Can I view the current UV spectrum in real-time mode and then get it back at any time?
- Is there a version that meets the requirements of routine work (including templates, GLP-standard documentation, supervisor routine) as well as a developer version (flexibility in changing method parameters)?
- Is software available for LC, LC-MS, LC-DAD and perhaps GC?
- How user-friendly is the software? Are there convenient export/import functions catering to a variety of formats? Can reduced data such as chromatograms, calibra-

tion graphs, etc., be integrated into mainstream word processing and statistics programs?
- What do I make of the frequent updates of the program? Are they a sign of hyperactivity or real innovation on the manufacturer's side?
- Has the software validation been carried out in a practice-relevant way?

Trends

Finally, I would like to draw your attention to 2–3 developments that I personally find remarkable and may give some idea of what the future holds.

1. The opportunities the Internet offers are more widely and more purposefully used. Examples:

- An increasing number of firms offer web-based control software.
- Intelligent filters help you find colleagues that are dealing with just the same separation problems you are trying to solve.
- Freshly generated MS spectra can be compared with spectra from spectral libraries within seconds.

2. Straightforward, well thought out solutions protect the environment and enhance flexibility, as the following examples show:

- "Environment-responsive chromatography" – a functional group at the surface of a stationary phase shows hydrophobic properties at elevated temperature, whereas the same ligand is hydrophilic at lower temperature. Thus, the selectivity can be controlled just via the temperature.
- There are several functional groups incorporated in the stationary phase, which means that the same column can be used in NP as well as in RP mode, depending on the eluent chosen. This helps reduce the number of columns needed for different applications.

3. Two trends can be observed in the pharmaceutical industry – the emphasis on strict formal requirements on the one hand, and more customer-friendliness and pragmatism on the other. This can be illustrated by three examples:

- In purity profiles, it will soon be a legal requirement to separate all potentially occurring impurities. By contrast, in stability indicating assays, demands are slightly more modest – only all possible degradation products (i.e., not all peaks) have to be separated. This could be achieved using two different methods. A system suitability test is to be carried out and documented exclusively for the reporting limit.
- As long as requirements – e.g., a certain resolution – are met, "adjustments" are handled quite generously. "Change control" or even revalidation can be reduced to a minimum by using "intelligent wording" in the validation report.
- A company that has an impurity to report for a product that has been licensed in the USA, Europe and Japan will have to do so following three validated (!) methods – in order to meet the USP, EP and JP regulatory standards.

3.6 Thoughts About a Dead Horse

Dear Reader,

What I am trying to do in my trilogy of HPLC tips – Volume 3 will be published soon – is to approach everyday problems in HPLC rationally. This may go against the grain of the Zeitgeist. Reason is up against what is called logic. We live in an age where contents, basic insights and long-term strategies don't seem to matter. Excel acrobats, rubber stamp fetishists, packaging whiz kids, label designers and experts in surface treatment rule supreme, be it in the chemical or pharmaceutical industry or elsewhere.

It is quite likely that in the long run, this mindset will be eroded from within and collapse, but until then there is nothing left but to grin and bear it. I think this note, which I found on the pinboard of a pharmaceutical firm, will help us do just that.

A saying of the Dakota Indians goes:
"If You Find Yourself Riding a Dead Horse, Get Off"

In our professional life, however, when confronted with such a situation, we would often follow different strategies:

1. Get a stronger whip.
2. Change riders.
3. Say "That's the way we have always ridden our horses".
4. Found a committee that analyses the horse.
5. Go elsewhere to learn how dead horses are ridden there.
6. Raise our quality standards for riding dead horses.
7. Create a task force to resuscitate the dead horse.
8. Go on a training course to improve our riding skills.
9. Initiate a comparative study of dead horses.
10. Change the criteria that define when a horse is dead.
11. Hire new people to ride the dead horse.
12. Harness several dead horses together to speed them up.
13. Proclaim that no horse is so dead that it cannot be whipped.
14. Find additional funds to raise the performance of the horse.
15. Initiate a study to reduce consulting costs.
16. Buy a tool that promises to make dead horses run faster.
17. Claim that our horse is dead in a better, faster and cheaper way.
18. Found a quality circle investigating how dead horses can be put to good use.
19. Revise the operating conditions for dead horses.
20. Introduce independent auditing for dead horses.

Index

100% method (normalized area method) 233, 244

acetonitril vs. methanol 180
acids, separation of 24, 27, 51
addition / additional
– method 233, 247 ff.
– peaks 103 ff.
additives (modifier)
– to the eluent 66 ff.
– in LC-MS 195 ff.
adduct
– alkali in LC-MS 208
– formation 197 ff.
aerosol 192 f.
– alkali cations in 197
aging of polar stationary phases 11 ff.
air bubbles 84, 101, 105, 168
algae 132
ammonium phosphate vs. potassium phosphate 33
analysis time, reducing of 55 f.
antidepressants, tricyclic 24, 63
antidote 198
APCI (Atmospheric Pressure Chemical Ionisation) 189 ff.
– linearity of APCI methods 193
API (Atmospheric Pressure Ionisation) 189 ff.
APPI (Atmospheric Pressure Photoionisation) 189 ff.
aspiration rate 168
Atmospheric Pressure (AP) 189 ff.
– Atmospheric Pressure Chemical Ionisation (APCI) 189 ff.
– Atmospheric Pressure Ionisation (API) 189 ff.
– Atmospheric Pressure Photoionisation (APPI) 189 ff.

bacteria 132
band broadening, contribution of the individual modules to 146 ff.
basic compounds
– selectivity of 19 ff.
– separation of 51 ff.
– – in the alkaline medium 49 ff.
– tailing by 116
breakthrough time (dead time) 214
buffers
– capacity (ion strength) 34, 112
– in common HPLC 35 ff.
– in LC-MS 195 ff.
bunching rate 238

calibration 237, 247
– curves 239
– – in LC-MS 206
catalytic effect
– of the metal syringe 103
– of the silica gel 94
cell
– impact of 216 f.
– path length 218
CID (collision-induced dissociation) 209
coefficient of variation 237 ff.
collision-induced dissociation (CID) 209
column / column parameters
– dual columns 89
– diameter and eluent consumption 276
– dimensions

More Practical Problem Solving in HPLC. S. Kromidas
Copyright © 2005 WILEY-VCH Verlag GmbH & Co. KGaA, Weinheim
ISBN: 3-527-31113-0

– – and gradient separation   159 ff.
– – by gradient runs   155
– guard   223
– lifetime of   141
– optimization of   14
– oven   134
– overloading of   115
– post-column addition in LC-MS   200
– RP-columns, selectivity of   31 f.
– saturation   223
column-selection   75 f.
column-switching valve   75 f., 220
conductivity detector   175
Coulomb explosions   190
coupling   90, 286 f.
cut-off of buffer solutions   34

dead
– time (breakthrough time)   214
– volume   65, 114, 144, 166
– – in µ-LC   214
detection limit, optimization via injection volume   164 f.
dopants   193
double peaks   91 f.
drift, elimination of, in gradient runs   96 f.
dual columns   89

efficiency
– increasing   63 ff.
– lower   214 ff.
electrochemical detectors   102
Electrospray Ionisation (ESI)   189 ff.
eluent consumption, and column diameter   276
elution order (reversed elution)   35, 37, 42, 46, 91
– change of   110 ff.
embedded phases   24
enrichment of the sample   220
environment-responsive chromatography   289
equilibration time in a gradient   270
equilibrium
– acid-base   195
– silanol groups   40

ESI (Electrospray Ionisation)   189 ff.
EXCEL
– calculation of the coefficient of variation   237
– calculation of a linear regression   245
– weighted regression with   251
excesses by assay analysis   131
exclusion   183
external standard method   231, 244 ff.

flow
– in gradient runs   60 f., 155
– in isocratic runs   55 ff.
– in LC-MS   189 ff.
– rate limit   227
fluorescence detector   175
fragment   208
fronting   119 ff.
F-test   249
fungi   132
fused silica   219

ghost peaks   94 f., 103 ff., 105 f., 107
gradient
– accuracy of   224 ff.
– delay volume in µ-LC   224
– duration, dependance of resolution on   275
– runs
– – correction in µ-LC   226
– – optimization of   60 f., 155 f.
– separation, and column dimensions   159 ff.
guard column   223

HPLC
– buffers in common HPLC   35 ff.
– parameters of an HPLC instrument   167 ff.
– software for   287 ff.
– websites for analytical chemistry   277 ff.
hydrophilic endcapping   24

injection
– systems in µ-LC   222
– volume   81, 215, 272 ff.

integration
- conditions 237 f.
- parameters, effect on peak area and peak hight 241 ff.
interfaces, LC-MS 189 ff.
internal standard method 232, 245 ff.
interstitial retention time 183
ion
- molecular ion 190 ff.
- paired ions in LC-MS 189 ff.
- source 190 ff.
- strength (buffer capacity) 34, 112
- suppression 208
isomers, separation of 21

knots in capillaries 171 ff.

LC (Liquid Chromatography)
- micro-LC 213 ff.
- nano-LC 213 ff.
LC-MS (Liquid Chromatography – Mass Spectrometry)
- additives in 195 ff.
- adduct alkali in 208
- buffers in 195 ff.
- calibration curves in 206
- coupling 187 ff.
- flow in 189 ff.
- interfaces in 189 ff.
- linear response in 205
- paired ions in 189 ff.
- pH-value in 192, 196
- post-column addition in 200
- sensitivity, enhance in 203
- thermolability in 194
lifetime of the column 141
limit of quantification 236 ff., 251
linear response in LC-MS 205
linearity of APCI methods 193

mass
- range 191
- spectrometer 189 ff.
memory effect 105, 177
methanol vs. acetonitril 180
method develoment in RP chromatography 74 ff., 78 ff.

micro-LC 213 ff.
microspray 191
miniaturization 152 f.
modifier (additives) 66 ff., 195 f.
molecular ion 190 ff.
monoliths 285
MS-MS 208

nano-LC 213 ff.
nanospray 191
nebulizer 190, 203
nitrile phase 29
nitrogen 190, 204
non-endcapped phases 25
normalized area method (100% method) 233, 244

orifice 190
orthogonal conditions 82
overlapping peaks, quantification of 229
overloading 215
- of the column 115

paired ions in LC-MS 189 ff.
parameters of an HPLC instrument 167 ff.
peak
- additional peaks 103 ff.
- area
- - change of peak area 127, 129 f.
- - effect on integration parameters 241 ff.
- - impact on peak area 230
- - quantification using 229 f.
- capacity 55
- double peaks 91 f.
- ghost peaks 94 f., 103 ff., 105 f., 107
- hight
- - change of peak hight 129 f.
- - effect on integration parameters 241 ff.
- - quantification using 229 f.
- overlapping peaks, quantification of 229
- small peaks, problems with 163
- width 238, 243

293

pH-value
– the dependance of UV-absorption 47 f.
– in LC-MS 192, 196
– the role of 40
– shift of 42 ff., 46 ff.
plasma 192, 195, 203
polar C_{18} phases, selectivity of 23
post-column addition in LC-MS 200
potassium phosphate vs. ammonium phosphate 33
precursor 208

quantification, limit of 236 ff., 251

refraction detector 175
regression, linear 244 ff.
resolution 14, 161 f., 270
response time/time constant 168, 243
restrictor capillary 84
retention factor 270
reversed elution (*see* elution order) 35, 37, 42, 46, 91
rise time/time constant 168, 243
robustness 135 ff.
RP-chromatography, method develoment in 74 ff., 78 ff.
RP-columns, selectivity of 31 f.

sampling
– rate 168, 238, 244
– time 238, 241, 244
saturation column 223
selectivity 161 f.
sensitivity

– enhance in LC-MS 203
– gain in 218
separation factor 162, 270
silica, fused 219
single reaction monitoring 208
slit width 168
slope 238, 244
small peaks, problems with 163
soft clipping 205
software for HPLC 287 ff.
standard deviation 251
– relative 256
standard method
– external 231, 244 ff.
– internal 232, 245 ff.
steric aspects, impact on resolution 17
steroids 25 f.
student's value t 236
suppression, integration 238

F-test 249
thermolability in LC-MS 194
threshold 238, 244
time constant 168, 243

uncertainty 250

variance 146
variation, coefficient of 237 ff.

wavelength, the right 171 ff.
websites for analytical chemistry, HPLC 277 ff.
weighted regression 249 ff.
wettability 108